Electrical Applications for Air Conditioning & Refrigeration Systems

by
Billy C. Langley

Electrical Applications for Air Conditioning & Refrigeration Systems

by
Billy C. Langley

Published by
THE FAIRMONT PRESS, INC.
700 Indian Trail
Lilburn, GA 30047

Library of Congress Cataloging-in-Publication Data

Langley, Billy C., 1931-
Electrical applications for air conditioning & refrigeration systems
by Billy C. Langley.
    p. cm.
    Includes index.
    ISBN 0-88173-273-7
    1. Air conditioning—Electric equipment. 2. Refrigeration and
refrigerating machinery—Electric equipment. I. Title.
TK4035.A35L36   1999   697.9'3--dc21   98-54804
                                                    CIP

Published by The Fairmont Press, Inc.
700 Indian Trail
Lilburn, GA 30047

Printed in the United States of America

10 9 8 7 6 5 4 3 2 1

ISBN 0-88173-273-7 FP

ISBN 0-13-014307-3 PH

While every effort is made to provide dependable information, the publisher, authors, and
editors cannot be held responsible for any errors or omissions.

Distributed by Prentice Hall PTR
Prentice-Hall, Inc.
A Simon & Schuster Company
Upper Saddle River, NJ 07458

Prentice-Hall International (UK) Limited, London
Prentice-Hall of Australia Pty. Limited, Sydney
Prentice-Hall Canada Inc., Toronto
Prentice-Hall Hispanoamericana, S.A., Mexico
Prentice-Hall of India Private Limited, New Delhi
Prentice-Hall of Japan, Inc., Tokyo
Simon & Schuster Asia Pte. Ltd., Singapore
Editora Prentice-Hall do Brasil, Ltda., Rio de Janeiro

Preface

Electrical Applications for Air Conditioning & Refrigeration Systems

Electrical Applications for Air Conditioning & Refrigeration Systems is designed in units so that the reader will only need to refer to the section to which he/she is concerned. This eliminates the need for thumbing through the complete book for the wanted information.

Each unit is separated into a presentation of that particular subject, control or operational theory. Also included is troubleshooting procedures and check-out procedures so the reader will have the necessary information at hand for quick reference. The reader may also want to study the material at a more leisurely time so that a better understanding is obtained from the material presented.

The material covers basic electricity in an easy to follow format. Electronics is presented so that the reader can learn the basics of solid-state theory or just study the material required for the particular job at hand. The controls are presented in a clear and concise manner so the reader can quickly learn how the control works, how it is used in the system. Information on how to make operational checks and troubleshoot the control for proper operation are also included.

There are ninety-one units in the book. Each one is devoted to the control, theory, or troubleshooting procedure for that particular component. Electrical Applications for Air Conditioning & Refrigeration Systems will make a good reference book as well as a self study course for anyone needing instruction in this field of study.

Contents

Electrical Applications for Air Conditioning & Refrigeration Systems

SECTION 1: ELECTRICAL FUNDAMENTALS

Electricity is probably the most important form of energy used in the world today. Because of electricity we have electric lighting, television, radio, telephone service, and automobile operation. With new developments almost daily, and new applications electricity offers virtually unlimited possibilities for the technician who will put forth the effort to learn the theories and applications.

INTRODUCTION

Approximately 75 to 80% of the problems encountered in air conditioning and refrigeration work is due to some electrical or electronic malfunction. The technician who knows and can readily apply the theories involved will have a much easier and rewarding carrier in this exciting industry.

Electrical troubleshooting is most likely the most neglected area of maintaining air conditioning and refrigeration equipment. Troubleshooting the electrical or electronic circuits of these systems is fairly simple when the theories are understood and applied. Each and every component in and electrical circuit must either be in the circuit or have some means to cause it to become a part of the complete circuit. Also, the components within an electrical circuit must consume all of the electrical energy applied to it. Only when these two basic needs are met will electro-mechanical work be done.

Unit 1: What Is Electricity?

We are all aware of the effects of electricity. It has been a vital part of our everyday lives and yet very few of us can give a satisfactory explanation as to what electricity actually is. For years people have known how to produce it, transport it, and use it; but few actually knew exactly what it was. Today, however, we know that, for practical purposes, it is produced by the movement of small atomic particles known as electrons. The movement of these electrons through conductors (wires) is what causes the effects of electricity as we know it.

The force (voltage) that causes these small atomic particles to move though the conductor, the effects produced when opposition (resistance) is met, and how these different forces can be used to effectively control the flow of electricity are what is known as the basic principles of electrical theory. In actuality, these atomic particles cannot be seen by the naked eye; however, they are present in all matter. Before a complete understanding of how these particles exist, the structure of matter must first be understood.

Unit 2: ATOMIC THEORY:

The atomic theory states that all matter has an electrical structure, and is therefore, electrical in nature. All objects are made up of a combination of positive and negative electrical charges. Therefore, electrical current will flow through everything: your body, a long wire, or through a stream of impure water. It is easier to create this flow of electricity through some substances than it is through others. It can be seen that any study of electricity must begin with a study of the structure of matter.

MATTER: Matter is defined as any substance that has mass and occupies space. It can be felt, seen, or used. Mass may be in almost any form; such as: a solid, liquid, or gas (vapor). Some of the more popular solids are wool and metal. Water and gasoline are forms of a liquid. Oxygen and hydrogen are popular forms of a gas.

Structure of Matter: The definition of matter given previously is broad enough to include all the objects in the universe. Do not confuse energy with matter. Even though they are closely related, energy does not have mass nor does it occupy space.

The Elements: Elements are considered to be the basic building blocks that make up all matter. Some of the more common elements are oxygen, hydrogen, silver, and gold. There are approximately 105 elements in existence.

Everything about us is made from the elements. However, it should be noted that the actual elements themselves cannot be produced or reduced by simple chemical processes.

The Compound: Some types of materials will form a chemical union when combined with one another in the proper proportions. When such a union occurs, the new substance, which is different from either of the elements used to form it, is called a compound.

Actually, there are many more materials than there are

elements to form them. This occurs because the different elements can be combined to produce materials having characteristics that are completely different from the elements.

Example: Hydrogen and oxygen are two gaseous elements, which, when combined in the proper quantities combine into a liquid commonly known as water. See Figure 2-1.

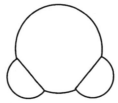

Fig. 2-1 A molecule of water.

The Molecule: When a substance has been broken down into the smallest particle resembling the original substance the remaining particle is known as the molecule. In our previous example of water, water contains two atoms of hydrogen and one atom of oxygen to form the chemical union. See Figure 2-2.

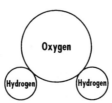

Figure 2-2 Hydrogen and Oxygen Atoms.

As another example, if we were to take a crystal of table salt and keep dividing it into half we would still have a molecule of salt. If we split the molecule further, the salt would return to its original elements. See Figure 2-3.

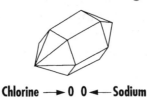

Chlorine ➝ 0 0 ◂— Sodium

Figure 2-3 Salt Elements.

The Atom: Electrical theories are based on the atom as being the smallest particle of an original substance. When water is reduced to its smallest size, we would have one molecule of water. When the water molecule is reduced, we would have hydrogen and oxygen atoms. See Figure 2-4.

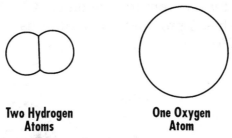

Two Hydrogen Atoms　　　　**One Oxygen Atom**

Figure 2-4 Water broken down into atoms.

Structure of the Atom: When we break down the atom of an element further, the original element would not be present in the remaining particles. This is because these smaller particles or matter are present in all the atoms of different elements. The atom of one element is different from the atom of a different element. This is because each element contains different amounts of these subatomic particles.

There are three of these subatomic particles that are of interest in the study of electricity. These particles are known as: protons, neutrons, and electrons. The protons and neutrons are located in the nucleus, or center, of the atom. The electrons travel around the nucleus in paths called orbits. See Figure 2-5.

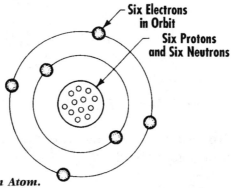

Six Electrons in Orbit

Six Protons and Six Neutrons

Figure 2-5 Parts of an Atom.

The Nucleus: The nucleus is located in the center of the atom. It is made up of protons, neutrons, and other subatomic particles. The nucleus has exactly the same electrical charge as an electron, but it is opposite in value, or sign. The number of protons in the nucleus determines how one atom of an element is different from another atom. Some examples are: Helium has 2 protons, magnesium has 12 protons, and uranium has 92 protons. in the nucleus. See Figure 2-6.

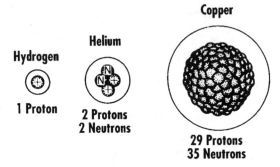

Figure 2-6 Protons in the nucleus.

Even though the neutron is and individual particle, it is usually thought of as being a combination of electrons and protons making it electrically neutral. Because they are electrically neutral, neutrons are not very important in electrical theory. The nucleus of an atom contains neutrons, which have no electrical charge; and protons, which are positively charged; the nucleus of any atom is considered to always have a positive electrical charge. See Figure 2-7.

Figure 2-7 Electrical charges of a nucleus.

The Proton: The nucleus, or center, of an atom is made up mostly of positively charged particles called protons, and neutral particles called neutrons. Most of the weight of an atom is in the nucleus. It is estimated that the diameter of an proton is 0.07 trillionth of an inch. It is approximately one-third of the diameter of an electron, but it weighs approximately 1845 times more than an electron. It is very difficult to separate a proton from the nucleus of an atom. In electrical theory, therefore, they are considered to be permanent parts of the nucleus. Thus, protons have nothing to do with the flow of electrical current through a circuit.

Since the proton has a positive electrical charge the lines of force caused by the flow of electricity through a conductor will extend directly out from the proton in all directions. See Figure 2-8.

Figure 2-8 Lines of force from a proton.

The Electron: Electrons are particles of electricity that have a negative electrical charge and revolve around the nucleus in orbits. Centrifugal force prevents these electrons from falling to the center of the atom. Usually, there is one electron in an orbit for each proton in the nucleus. The positive charges of the protons and the negative charges of the electrons cause the atom to be electrically neutral. The atomic structure of each element could be described as having a fixed number of electrons in orbit around the nucleus. Some examples of the atomic structure of different elements are shown in Figure 2-9.

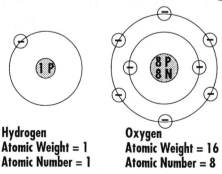

Hydrogen
Atomic Weight = 1
Atomic Number = 1

Oxygen
Atomic Weight = 16
Atomic Number = 8

Figure 2-9 Atomic structure of hydrogen and oxygen.

The electron is approximately three times larger in diameter than and weighs about 1845 times less than the proton which makes it relatively easy to move. Thus, it can be seen that electrons are the particles that cause the flow of electrical energy through the circuit. The lines of force of the charge on electrons come straight in to the electron from all sides. See Figure 2-10.

Figure 2-10 Direction of lines of force for an electron.

Behavior of Electrons: Sometimes when we walk across a wool rug or slide across a car seat, electrons are displaced from one material to another. Then when we touch another object the electrons attempt to equalize themselves and an electrical spark is seen during this transfer of electrons. The same type of electrical charge can be generated by combing dry hair with a plastic comb. The comb will take on an excess of electrons which will cause lightweight objects to be attracted to it.

The electricity generated by rubbing two objects together is known as static electricity. The term static electricity is used because the electrons are collected on one of the materials and the electrical charge will remain on that material until it comes into contact with another type of material having an opposite electrical charge.

Unit 3: Electrical Charges

The value of the electrical charge on a body indicates the amount of energy that body has. Whether the charge is positive (+) negative (-) will depend on the number of electrons on that body. When more than the normal amount are present the charge will be negative in nature. When fewer than the normal amount are present the body will have a positive charge.

UNCHARGED AND CHARGED ATOMS: When an atom has no charge it has an equal number of electrons and protons. See Figure 3-1.

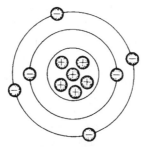

Figure 3-1 An uncharged atom.

When one or more electrons are removed from it, the atom becomes positively charged. See Figure 3-2.

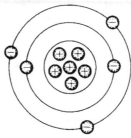

Figure 3-2 A charged atom.

When an electron is forced from its atom, it will cling to anything close to it. This action causes the atom from where it came to be positively charged and the atom to which it clings to be negatively charged. When the atom with an excess of electrons moves close to one that has an excess of protons, the excess electrons will move from the negatively charged atom to the positively charged one. This is done in an attempt to equalize the charges on both atoms. This transfer of electrons is usually accompanied by a spark or cracking sound.

LAWS OF ELECTRICAL CHARGES: When an particle is in the balanced condition, the negative electrical charge of the electron is equal to the amount of positive charge on the proton. These charges are known as electrostatic charges. The lines of force associated with each particle produce electrostatic fields around the particle. The way that these fields react with each other will cause charged particles to either repel or attract each other. The law of electrical charges states that: particles with like charges will repel each other, and those with unlike charges will attract each other. See Figure 3-3.

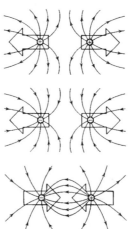

Figure 3-3 Law of electrical charges.

Thus: An electron (–) will repel other electrons (–) = (Like charges repel.)

A proton (+) will repel another proton (+) = (like charges repel.)

A proton (+) will attract an electron (–) = (unlike

charges attract.)

Because the protons are relatively heavy, they have very little repulsive affect on each other in the nucleus.

IONIZATION: An atom almost always remains in its normal state unless it receives energy from some external source such as friction, heat, or from being bombarded by electrons from another atom. When this additional energy is added to the atom it will become excited, or charged, and the loosely bound electrons in the outer orbits may leave the atom. When electrons leave an balanced atom there will be a deficiency of electrons and the atom will no longer be neutral in charge. The atom is then said to be ionized. When electrons are lost the atom becomes a positive ion. When electrons are gained the atom becomes a negative ion.

ELECTROSTATIC FIELDS: The field of force that surrounds a charged body is known as an electrostatic field. The attracting and repelling forces around charged objects are present because of the electrostatic lines of force present around all charged bodies.

The electrostatic lines of force surrounding a negatively charged body come into the object from all directions. The lines of force surrounding a positively charged body go out from the object in all directions. See Figure 3-4.

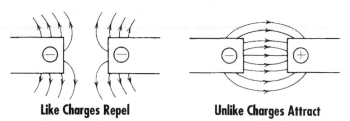

Like Charges Repel   **Unlike Charges Attract**

Figure 3-4 Electrostatic lines of force.

These lines of force either strengthen or weaken the electrostatic field depending on whether they aid or oppose each other.

The two factors that determine the attraction or repulsion strength of electrostatic fields are: (1) the amount of charge on each body, and (2) the distance between to two bodies. A larger electric charge on each body will produce more lines of force than a smaller electric charge. Thus, a greater electrostatic force will be present. When the two bodies are close a stronger electrostatic field will be produced. This is because the lines of force are more concentrated closer to the charged body.

ATTRACTION AND REPULSION: If we positively charge a glass rod by rubbing it with a piece of silk, and negatively charge a rubber rod by rubbing it with a piece of fur and prevent them from touching anything, we can see the effects of attraction and repulsion. See Figure 3-5.

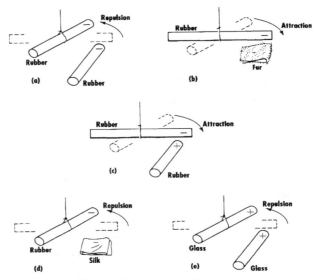

Figure 3-5 Attraction and repulsion examples.

The rubber rod is suspended so that it can move freely, then bring another rubber rod with he same charge close to it but without them touching, the suspended rod will move away. See Figure 3-5a.

When the piece of fur is brought close to the suspended rubber rod without them touching, the opposite charges will cause the two objects to be attracted to each other. The rod will tend to swing toward the fur. See Figure 3-5b.

When the positively charged glass rod is brought close to the suspended rubber rod without them touching, the two rods will be attracted to each other. Causing the suspended rubber rod to swing toward the glass rod. See Figure 3-5c.

When a piece of silk is brought close to the suspended rubber rod without them touching, the like charges will tend to cause the rubber rod to swing away from the piece of silk. See Figure 3-5d.

Another experiment that can be performed by suspending a positively charged glass rod with a string so that it is free to move. Then bring another positively charged glass rod close to the suspended glass rod, without them touching, the suspended glass rod will tend to move away from the more stationary glass rod, thus proving that like charges repel and unlike charges attract. See figure 3-5e.

Unit 4: Electron Flow

Electricity is produced when electrons leave their atom because the atoms become charged ions. There are several ways of causing these electrons to leave their orbits around the nucleus of an atom. To understand how this is possible, we must know something about the nature of the electron orbits.

ELECTRON ORBITS: The electrons, in their orbital path around the nucleus move at a high rate of speed. The centrifugal force caused by this high rate of speed tends to cause the electron to leave its orbit, or shell, and fly off into space. However, the attraction of the nucleus, because of its positive charge, tends to keep the electron from leaving its shell. See Figure 4-1.

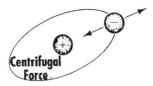

Figure 4-1 Normal forces on a revolving electron.

When enough outside force is applied to the electron to help the centrifugal force, it can be forced out of its orbit and become a "freed" electron. See Figure 4-2.

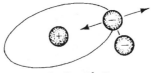

Figure 4-2 Outside force on an electron.

ORBITAL SHELLS: Electrons are located in the atomic shells that surround the nucleus of every atom. Each atom can have up to seven shells, depending on the nature of the element. See Table 4-1.

Table 4-1 Electron Shells

Atomic No.	Element	Electrons per shell				
		1	2	3	4	5
1	Hydrogen, H	1				
2	Helium, He	2				
3	Lithium, Li	2	1			
4	Beryllium, Be	2	2			
5	Boron, B	2	3			
6	Carbon, C	2	4			
7	Nitrogen, N	2	5			
8	Oxygen, O	2	6			
9	Fluorine, F	2	7			
10	Neon, Ne	2	8			
11	Sodium, Na	2	8	1		
12	Magnesium, Mg	2	8	2		
13	Aluminum, Al	2	8	3		
14	Silicon, Si	2	8	4		
15	Phosphorus, P	2	8	5		
16	Sulfur, S	2	8	6		
17	Chlorine, Cl	2	8	7		
18	Argon, A	2	8	8		
19	Potassium, K	2	8	8	1	
20	Calcium, Ca	2	8	8	2	
21	Scandium, Sc	2	8	9	2	
22	Titanium, Ti	2	8	10	2	
23	Vanadium, V	2	8	11	2	
24	Chromium, Cr	2	8	13	1	
25	Manganese, Mn	2	8	13	2	
26	Iron, Fe	2	8	14	2	
27	Cobalt, Co	2	8	15	2	
28	Nickel, Ni	2	8	16	2	
29	Copper, Cu	2	8	18	1	
30	Zinc, Zn	2	8	18	2	
31	Gallium, Ga	2	8	18	3	
32	Germanium, Ge	2	8	18	4	
33	Arsenic, As	2	8	18	5	
34	Selenium, Se	2	8	18	6	
35	Bromine, Br	2	8	18	7	
36	Krypton, Kr	2	8	18	8	
37	Rubidium, Rb	2	8	18	8	1
38	Strontium, Sr	2	8	18	8	2
39	Yttrium, Y	2	8	18	9	2

Atomic No.	Element	Electrons per shell						
		1	2	3	4	5	6	7
40	Zirconium, Zr	2	8	18	10	2		
41	Niobium, Nb	2	8	18	12	1		
42	Molybdenum, Mo	2	8	18	13	1		
43	Technetium, Tc	2	8	18	14	1		
44	Ruthenium, Ru	2	8	18	15	1		
45	Rhodium, Rh	2	8	18	16	1		
46	Palladium, Pd	2	8	18	18	0		
47	Silver, Ag	2	8	18	18	1		
48	Cadmium, Cd	2	8	18	18	2		
49	Indium, In	2	8	18	18	3		
50	Tin, Sn	2	8	18	18	4		
51	Antimony, Sb	2	8	18	18	5		
52	Tellurium, Te	2	8	18	18	6		
53	Iodine, I	2	8	18	18	7		
54	Xenon, Xe	2	8	18	18	8		
55	Cesium, Cs	2	8	18	18	8	1	
56	Barium, Ba	2	18	18	18	8	2	
57	Lanthanum, La	2	8	18	18	9	2	
58	Cerium, Ce	2	8	18	19	9	2	
59	Praseodymium, Pr	2	8	19	20	9	2	
60	Neodymium, Nd	2	8	19	21	9	2	
61	Promethium, Pm	2	8	18	22	9	2	
62	Somarium, Sm	2	8	18	23	9	2	
63	Europium, Eu	2	8	18	24	9	2	
64	Gadolinium, Gd	2	8	18	25	9	2	
65	Terbium, Tb	2	8	18	26	9	2	
66	Dysprosium, Dy	2	8	18	27	9	2	
67	HOlmium, Ho	2	8	18	28	9	2	
68	Erbium, Er	2	8	18	29	9	2	
69	Thulium, Tm	2	8	18	30	9	2	
70	Ytterbium, Yb	2	8	18	31	9	2	
71	Lutetium, Lu	2	8	18	32	9	2	
72	Hafnium, Hf	2	8	18	32	10	2	
73	Tantalum, Ta	2	8	18	32	11	2	
74	Tungsten, W	2	8	18	32	12	2	
75	Rhenium, Re	2	8	16	32	13	2	
76	Osmium, Os	2	8	18	32	14	2	
77	Iridium, Ir	2	8	18	32	15	2	
78	Platinum, Pt	2	8	18	32	16	2	

Table 4-1 Electron Shells

Atomic No.	Element	Electrons per shell							Atomic No.	Element	Electrons per shell						
		1	2	3	4	5	6	7			1	2	3	4	5	6	7
79	Gold, Au	2	8	18	32	18	1		92	Uranium, U	2	8	18	32	21	9	2
80	Mercury, Hg	2	8	18	32	18	2		93	Neptunium, Np	2	8	18	32	22	9	2
81	Thallium, Tl	2	8	18	32	18	3		94	Plutonium, Pu	2	8	18	32	23	9	2
82	Lead, Pb	2	8	18	32	18	4		95	Americium, Am	2	8	18	32	24	9	2
83	Bismuth, Bi	2	8	18	32	18	5		96	Curium, Cm	2	8	18	32	25	9	2
84	Polonium, Po	2	8	18	32	18	6		97	Berkelium, Bk	2	8	18	32	26	9	2
85	Astatine, At	2	8	18	32	18	7		98	Californium, Cf	2	8	18	32	27	9	2
86	Radon, Rn	2	8	18	32	18	8		99	Einsteinium, E	2	8	18	32	28	9	2
87	Francium, Fr	2	8	18	32	18	8	1	100	Fermium, Fm	2	8	18	32	29	9	2
88	Radium, Ra	2	8	18	32	18	8	2	101	Mendelevium, Mv	2	8	18	32	30	9	2
89	Actinium, Ac	2	8	18	32	18	9	2	102	Nobelium, No	2	8	18	32	31	9	2
90	Thorium, Th	2	8	18	32	19	9	2	103	Lawrencium, Lw	2	8	18	32	32	9	2
91	Protactinium, Pa	2	8	18	32	20	9	2									

The electrons located in shells closer to the nucleus are more difficult to free because of the strong attraction to the positive charge of the nucleus. However, when the electrons gain energy they will move into the next shell, which is further away from the nucleus. Their attraction to the nucleus then becomes weaker. As an electron moves to the next shell its attraction to the nucleus becomes less. This is because as the distance between two charges increases the attraction becomes weaker. The more electrons an atom has, the more shells there are in that atom.

Shell Capacity: Each shell of an atom of a given element can hold only a certain number of electrons. Notice in Table 4-1 that the first shell, the one closest to the nucleus, can hold only two electrons (column 1); the second shell can hold no more than 8 electrons (column 2); the third shell can hold no more than 18 (column 3); the fourth can hold no more than 32 (column 4); and so on. See Figure 4-3.

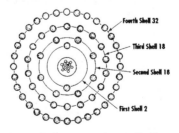

Figure 4-3 Maximum number of electrons in a shell.

The Valence Shell: The shell the farthest from then nucleus is called the valence shell. Therefore, the electrons located in this shell are known as the valence electrons. See Figure 4-4.

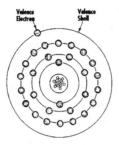

Figure 4-4 Location of valence electrons.

The electrons contained in this shell are important in the study of electricity. The smaller the number of valence electrons in an atom the easier it is to free them.

Note that even though the third shell can hold a maximum of 18 electrons (Refer to Table 4-1), it does not hold no more than 8 until the fourth shell is started. This is true for all the outer shells. Each shell after this will not take on more than 8 electrons regardless of how many it can hold. Thus, we know that the outer shell of an atom will never have more than 8 electrons in it.

Electron Energy: All electrons have a negative electrical charge; however, not all of them have the same amount of energy. The electrons located in the first shell (the shell closest to the nucleus) have less energy than those located in the outer shells. The electrons in the valence shell, (the shell farthest from the nucleus) have the most energy. See Figure 4-5.

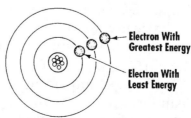

Figure 4-5 Energy in electrons in each shell.

As an electron gains in energy, it will move out of its orbit and go to the next higher orbit. When valence electrons gain enough energy they will move out of the atom. They will leave the atom because there are no other orbits for them to move into.

PRODUCING ELECTRICITY: Each and every time an electron leaves its atom electricity is produced. The valence electrons leave the atom because it takes less energy to free them because of their distance from the nucleus. The energy which is applied to the valence shell is divided evenly among all the electrons in it. Therefore, the more electrons in the shell the less energy each will receive. See Figure 4-6.

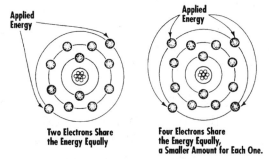

Two Electrons Share
the Energy Equally

Four Electrons Share
the Energy Equally,
a Smaller Amount for Each One.

Figure 4-6 Energy applied to valence electrons.

SOURCES OF ELECTRICITY: In our earlier experiments we, learned that a body might take on an electrical charge. The polarity of this charge was determined by whether there was an excess of electrons, or a deficiency of electrons, on the body causing a potential difference or an electromotive force when the electrons were redistributed. People have done a lot of research and experimenting trying to develop something that will develop this imbalance in electrical charge and electrical pressure.

There are basically six different ways to produce electricity: (1) friction, (2) chemical action, (3) mechanical action, (4) heat, (5) light, and (6) magnetism. In this study we will discuss the most important to us: friction, chemical, and magnetism.

Friction Energy to Electric Energy: An electrical charge may be created when two pieces of different types of material are rubbed together. It may also be produced when someone walks across a dry carpet. Your shoe soles are charged by being rubbed on the carpet. This charge is then transferred to you and will be transferred to another object that you might touch. This is what causes the shock, or spark, when a doorknob or light switch is touched. These charges, called static electricity, occur when one type of material transfers electrons to another type. See Figure 4-7.

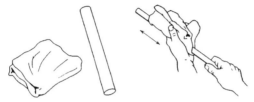

Figure 4-7 Static (Triboelectricity).

There is still some mystery around static electricity. However, one theory is: There are many atoms on the surface of a material that cannot combine with other atoms as they can inside the material. Thus, the surface atoms have free electrons. Because of this, insulators such as glass and rubber can produce static electricity. Any heat energy caused by the rubbing action is transferred to these surface atoms and causes them to release some electrons. This process is known as the triboelectric effect.

Chemical Energy to Electric Energy: Certain metals and chemicals can be combined to cause a transfer of electrons and therefore produce electrical charges. This process is known as the electrochemical principle. It may be illustrated in a description of the way in which an ordinary battery works. When sulfuric acid and water are mixed in a glass container, the sulfuric acid breaks down into hydrogen (H) and sulfate (SO_4). The hydrogen atoms have a positive charge and sulfate atoms have a negative charge. However, the number of positive and negative charges are equal and the entire solution has no potential difference.

When copper and zinc bars are added to the solution,

there is a reaction. The zinc and sulfate atoms combine. Because the zinc atoms are negatively charged, the zinc bar gives off positively charged ions. The electrons from the zinc ions are left in the zinc and, therefore, the zinc bar has a surplus of electrons causing it to be negatively charged. The zinc ions neutralize the sulfate ions and the solution now has a positive charge. The positive hydrogen ions attract free electrons from the copper bar and again neutralize the solution. Now the copper bar has a lack of electrons and the base is thus positively charged. See Figure 4-8.

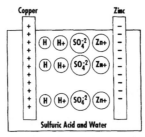

Figure 4-8 Electrochemistry action.

Magnetic Energy to Electric Energy: We have all experienced the effects of magnets. Because of this we know that sometimes two magnets will attract each other and sometimes they will repel each other. This attraction or repulsion force occurs because the magnets have force fields that interact with each other. See Figure 4-9.

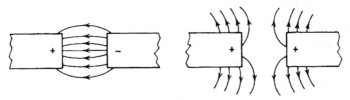

Figure 4-9 Interaction of magnetic fields.

It is also possible to move electrons with the force of a magnetic field. This type of electricity is called magneto-electricity. This is the basic principle used in generators to produce electricity. When a copper conductor is passed through a magnetic field, the force of the magnetic field will transfer enough energy to the copper to free the valence electrons from their atoms. Movement of the copper conductor in one direction will cause the electrons to flow in one direction, and moving it in the opposite direction will cause the electrons to flow in the opposite direction. See figure 4-10.

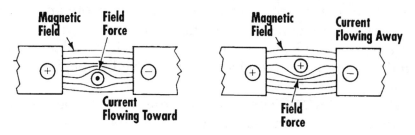

Figure 4-10 Electron flow caused by moving copper in a magnetic field.

The same affect can be observed when the copper conductor is held still and the magnetic field is moved. Actually, it is the relative motion between the conductor and the magnetic field that causes the electrons to flow.

STATIC ELECTRICITY: People have known about static electricity for any years. Static electricity is produced by permanently displacing electrons from their atom. Static electricity will provide no sustained flow of electrons, therefore; and electric current will occur for only a short period of time. When the charges between the two objects become equalized the flow of electricity stops. For electricity to be useful, there must be a constant flow of electrons available.

The static electricity that we are most familiar with is usually caused by friction. It is generally produced by walking across a carpet in very dry weather, by rubbing a piece of silk on a glass rod, or by rubbing a piece of fur on a rubber rod. Static electricity is simply a displacement of electrons from one substance to another.

Unit 5: Electric Current

Electric current is defined as the flow of free electrons through a circuit. Until now we have studied about static electricity, or electric charges at rest. It is necessary, however, to set these electric charges in motion before any useful work can be done. This is what happens when electric current is produced. A greater demand for electricity requires that more free electrons be moved through a conductor.

FREE ELECTRONS: For electron flow to be useful requires that many billion of them flow past a given point. They are caused to flow though a conductor by repelling each other because they have the same negative electrical charge. See Figure 5-1.

Conductor

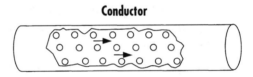

Figure 5-1 Free electrons flowing in the same direction through a conductor.

The direction of flow of the electrons is always from negative to positive. Any one single electron does not flow completely through a conductor. This electron movement may be compared to a long row of billiard balls being struck by a single ball. See figure 5-2.

Striking Force　　　　　　　　　　　　**Impulse Force**

Figure 5-2 Impulse transfer of energy.

This process is known as the impulse transfer of energy. When a moving ball strikes on on the end of the row, it bumps the next one in return. This process continues through the complete string of balls. Note that the last ball in the string moves at almost the same instant that the first one is struck.

Copper atoms have only one valence electron which is held loosely in its orbit. The atoms in a copper conductor are so close that their shells overlap. See Figure 5-3.

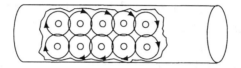

Figure 5-3 Random electron flow in a copper conductor.

The electrons moving around the nucleus can be easily moved to the orbit of another atom. This movement then causes an electron to be discharged from that atom, setting up a chain reaction in the remaining atoms. This electron movement is random and has no effect on the flow of electric current. The valance electrons are not really associated with any one particular atom. But, all the electrons are shared by all the atoms present.

ELECTRON MOVEMENT: For an electric current to exist, the electrons must be caused to flow in the same direction. If we connected one end of a copper conductor to a positive charge and the other end to a negative charge, such as an automobile battery, the electrons would be caused to move in the same direction because of the difference in the charges. See Figure 5-4.

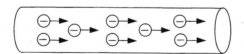

Figure 5-4 Electrons moving in the same direction.

The electrons, because of their negative charge, are repelled by the negative charge of the power source and attracted by the positive charge of the other side of the power source. Because this attraction and repulsion exists in the external power source, the electrons can move in only one direction, from negative to positive, and cause and electrical current to flow in that direction. When the applied charges are small, all of the electrons do not flow in a straight path through the conductor. However, when the strength of these charges is increased the electrons begin to flow in a straighter path and move faster through the conductor. Electric current flows at the rate of 186,000 miles per second--the speed of light.

FLOW OF CURRENT: Current is electrons flowing through an electric circuit. Any difference in charges can cause the electrons to flow. The most common power sources are the automobile battery and the alternating current generator that supplies the electricity to our homes and work places. See Figure 5-5.

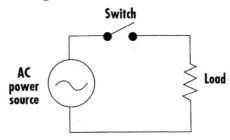

Figure 5-5 Basic electric system.

The amount of electrical potential (charge difference) and the amount of electrons flowing through the circuit are also shown in this figure.

Types of Current: There are two types of electric current used in HVAC&R equipment. They are direct current and alternating current. Direct current is either only positive or negative. It starts at a zero value and reaches a designed maximum and remains there until the circuit is opened. See Figure 5-6.

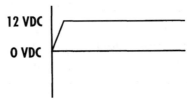

Figure 5-6 Direct Current (DC) voltage wave.

Direct current is used in many control circuits and in electronic ignition systems. As well as other uses in operating HVAC&R systems.

Alternating current is that supplied by the power company and is used to power the compressors and other types of motors and heating equipment in HVAC&R equipment. The sine wave for ac is half positive and half negative, hence the name alternating current. See Figure 5-7.

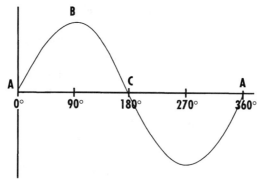

Figure 5-7 *Alternating current (AC) sine wave.*

Alternating current will be discussed in more detail in a later section.

VOLTAGE (POTENTIAL, EMF-ELECTROMOTIVE FORCE): Voltage is the electrical pressure or force that causes the electrons to flow though a closed circuit. Just like the water pressure causes water to flow through a pipe. It may be defined as the difference of electrical potential between two points. It may be designated by either of the letters E or V and the unit measured is the volt. The terms "potential," "electromotive force (EMF)," and "voltage" are commonly used interchangeably in the field. When a difference of potential causes 1 coulomb of current to do 1 joule of work, the EMF is 1 volt. Voltage is measured with a voltmeter. The units of voltage are shown in Table 5-1.

Table 5-1   Units of voltage

A charge of 1 coulomb = 6.28 × 10^{18} electrons
An emf of 1 volt (v) = 1 coulomb doing 1 joule of work
1 microvolt (μv) = 1/1,000,000 volt
1 millivolt (mv) = 1/1000 volt
1 kilovolt (kv) = 1000 volts
1 megavolt (megav) = 1,000,000 volts

Conversion of Units

volts (v) × 1000 = millivolts (mv)
volts (v) × 1,000,000 = microvolts (μv)
millivolts (mv) × 1000 = microvolts (μv)
volts (v) ÷ 1000 = kilovolts (kv)
volts (v) ÷ 1,000,000 = megavolts (megav)
megavolts (megav) × 1000 = kilovolts (kv)

millivolts (mv) ÷ 1000 = volts (v)
microvolts (μv) ÷ 1,000,000 = volts (v)
microvolts (μv) ÷ 1000 = millivolts (mv)
kilovolts (kv) × 1000 = volts (v)
megavolts × 1,000,000 = volts (v)
kilovolts ÷ 1000 = megavolts

1 v = 1000 mv = 1,000,000 μv = 0.001 kv = 0.000001 megav

CURRENT (AMPERE): The current flow through a circuit is measured in amperes, and is designated by (A), or (a). In Ohm's Law it is designated by (I). An ampere may be defined as the amount of current required to flow through a resistance of one ohm when a pressure of one volt is applied to it.

Current is the flow of electrons through a closed circuit. When electricity flows at the rate of 1 coulomb of electricity past a point in 1 second; a current of 1 ampere is flowing. The symbol for current in Ohm's Law is I. The units of current measurement are shown in Table 5-2.

Table 5-2 Units of current

1 ampere (a) = 1 coulomb/sec
1 milliampere (ma) = 1/1000 ampere
1 microampere (μa) = 1/1,000,000 ampere

Conversion of Units

amperes (a) × 1000 = milliamperes (ma)
amperes (a) × 1,000,000 = microamperes (μa)
milliamperes (ma) × 1000 = microamperes (μa)
milliamperes (ma) ÷ 1000 = amperes (a)
microamperes (μa) ÷ 1,000,000 = amperes (a)
microamperes (μa) ÷ 1000 = milliamperes (ma)

0.5a = 500 ma = 500,000 μa

Amperage is measured with an ammeter. There are versions of these instruments for both direct current and alternating current. These instruments are some of those used by installation and service technicians in the field to determine if a system is working as it should. When a current different from that recommended by the equipment manufacturer is detected, there is a problem.

RESISTANCE: Resistance may be defined as the opposition to current flow in an closed electric circuit. It should be remembered that anytime that current is flowing through a circuit there is some opposition to the movement of electrons. The unit of resistance is the ohm, and is designated by either I or R in most applications. When extremely small resistance is involved the microhm (Ω) which is one-millionth of an ohm is used. When larger resistance readings are involved the (MΩ) which is one million ohms or a megohm is used. Units of resistance are shown in Table 5-3.

Table 5-3   Units of resistance

ohms ÷ 1000 = kilohms (K)
ohms ÷ 1,000,000 = megohms (Meg)
kilohms (K) ÷ 1000 = megohms (Meg)
kilohms × 1000 = ohms
megohms × 1,000,000 = ohms
megohms × 1000 = kilohms (K)

500,000 ohms = 500 kilohms = 0.5 megohm

or

500,000 Ω = 500 K = 0.5 Meg

Table 5-3 Units of resistance.

A circuit has a resistance of one ohm when an EMF of 1 volt causes a current of 1 ampere to flow through it. Figure 5-8 shows two electric circuits with the same voltage but different resistances, and current flow. The different current flow is because of the difference in resistance.

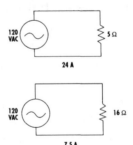

Figure 5-8 Current flow through a circuit with different resistances.

Usually, simple ohms are used in the air conditioning and refrigeration industry. Occasionally, there is a need for different readings such as the megohm (MΩ) which is one million ohms. These are generally used when testing the windings in electric motors.

All electric circuits and devices have some resistance that is inherent to them. The resistance of the components will depend on their designed use. Technicians in this field must know how this affects total unit operation and be able to make the necessary tests to determine the condition of unit components. When the resistance reading is outside the manufacturers recommended tolerances the component can be assumed defective and must be replaced.

Unit 6: Ohm's Law

The relationship between voltage **(E)**, current **(I)**, and resistance **(R)** in a closed circuit is known as Ohm's Law. This relationship may be expressed as follows: Current is directly proportional to voltage and inversely proportional to resistance. This can be simplified to:

1. If E is raised, I will go up.

2. If E is lowered, I will go down.

3. If R is raised, I will go down.

4. If R is lowered, I will go up.

The mathematical formula for this relationship is:

Current in amperes (I) = voltage in volts (E) ÷ resistance in ohms (R).

This equation can be transposed to find any missing value as long as the other two are known.

$I = E \div R$, $E = I \times R$, and $R = E \div I$

Ohm's Law triangle is often used as an aid in remembering these formulas. See Figure 6-1.

Figure 6-1 Ohm's Law triangle.

When the component to be solved for is known, simply cover that letter and the remaining portion of the equation will indicate the proper equation to use.

It should be remembered that Ohm's Law does not apply to alternating circuits because any coils in the circuit will produce some effects that will upset the law, but the general ideas presented do apply to alternating circuits. Alternating current will be discussed in a later chapter.

Example 1: What is the current flowing in the circuit in Figure 6-2.

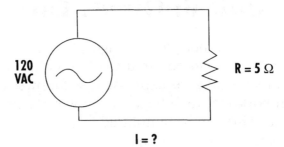

I = ?

Figure 6-2 Circuit for example 1.

Step 1 Determine which formula to use. I = E ÷ R.
Step 2 Fill in the known components: E = 120, R = 5 Ω.
Step 3 Solve the formula. I = 24 A.

Example 2: What is the resistance of a resistor if the voltage is 120 volts and the current is 0.75 amperes? See Figure 6-3.

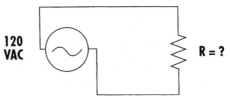

Figure 6-3 Circuit for example 2.

Step 1 Determine which formula to use. R = E ÷ I.
Step 2 Fill in the known components. E = 120, I = 0.75 A.
Step 3 Solve the formula. R = 160 Ω.

Example 3: What is the amount of voltage supplied to the circuit in Figure 6-4.

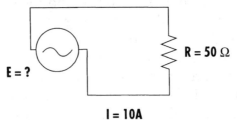

I = 10A

Figure 6-4 Circuit for example 3.

Step 1 Determine which formula to use. E = IR.
Step 2 Fill in the known components. I = 10A, R = 50Ω.
Step 3 Solve for the formula. E = 500 V.

CALCULATING ELECTRICAL POWER: The electric power that we pay for is measured in watts. The formula used is watts **(P)** = current **(I)** x volts **(E)**, or $P = E^2 \div R$, or $P = I^2R$. These letters can be placed in a device such as the Ohm's Law Triangle discussed earlier.

The following are some examples of calculating power in electric circuits.

Example 4: Determine the power used by an electric circuit that is supplied with 120 volts and is using 20 amperes. See Figure 6-5.

Step 1 Determine which formula to use. P = IE.

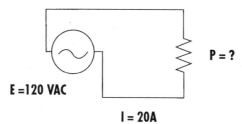

E =120 VAC

I = 20A

Figure 6-5 Circuit for example 4.

Step 2 Fill in the known components. P = 20 x 120.

Step 3 Solve for the formula. P = 2400 W.

Example 5: Determine the current draw of an electric heater that is rated for 1500 watts and is to be used on a 240 volt system. See Figure 6-6.

Step 1 Determine which formula to use. I = P ÷ E.

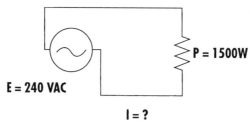

E = 240 VAC

I = ?

Figure 6-6 Circuit for example 5.

Step 2 Fill in the known components. I = 1500 ÷ 240.

Step 3 Solve for the formula. I = 6.25 A.

Example 6. Determine the power used in a circuit using 7 amperes current and having 15 ohms resistance. See Figure 6-7.

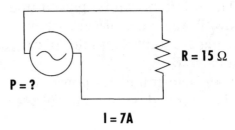

Figure 6-7 Circuit for example 6.

Step 1 Determine which formula to use. P = I²R.
Step 2 Fill in the known components. P = 7² x 15.
Step 3 Solve for the formula. P = 735 W.

Unit 7: Electrical Instruments

Air conditioning and refrigeration systems use electricity for control systems, fan motor operation, and compressor operation, to name the major uses of electricity for these systems. About 80% of all troubles with these systems are electrical in nature. To efficiently service these systems electric meters are used on a daily basis. It can be seen from this that a technician must be able to properly take readings and interpret these readings into useful information. There are several different types of meters used in the industry. Some are for special purposes while others are for general use.

During the installation of refrigeration and air conditioning systems, the installation technician must be able to read the necessary meters during the initial start up and testing of the system. This is almost a must so that future troubles can be eliminated before they actually become a problem.

During routine and emergency service procedures the service technician will need to read and interpret the readings of specific meters to determine the operating condition of the system. When meters are not used or their readings are not fully understood many problems will not be solved on the first attempt. This is expensive to the service company and an irritating problem to the equipment user.

The electrical characteristics on an electric circuit that are the most important to refrigeration, heating, and air conditioning technicians are voltage, amperage, and resistance. Electric meters are manufactured to measure either one or a combination which will include all three of these characteristics. The single use types are generally easier to use than the multiple use types. Also, when a multiple use meter must be repaired, some method must be devised to take the proper readings while the meter is in the repair shop. When a single use meter must be repaired, only theat type meter must be substituted rather than finding meters to replace the one that will read multiple characteristics.

ELECTRIC METERS: An electric meter is a device used to measure the characteristics of an electric circuit. The most popular types measure the voltage, amperage, and resistance of a circuit.

A basic analog type electric meter can be made with a solenoid coil and a soft-iron core to measure the electrical characteristics as it flows through a circuit. See Figure 7-1.

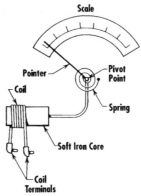

Figure 7-1 Basic electric meter movement.

As the current flows through the coil, a magnetic field is built up which draws the core into the coil. The strength of the magnetic field is in relation to the amount of current flowing through the wire. Thus, the core is drawn into the coil a distance related to the strength of the magnetic field. The opposite end of the core is fitted with a pointer and a spring. The pointer indicates the amount of current flowing through the wire and the spring tends to hold the core back, preventing its being drawn completely into the coil by a weak magnetic field. The scale usually has several different ranges. The proper range is selected by the user. To prevent damage to the meter, it is usually best to start with a range that is higher than the voltage or current that is suspected in the circuit. Then the proper scale can be chosen to get the most accurate reading.

Air conditioning and refrigeration technicians use meters to diagnose and help locate problems in an electric circuit. The technician must be able to accurately read the meter scales of all types of meters as quickly and efficiently as possible. These readings must be properly interpreted so that the problem can be quickly found and repaired.

When a new system has been installed or a major component replaced the technician should make a complete check of the electrical systems to make certain that everything is operating as it should. It should be remembered that no repair or installation is complete until there is a complete check of all electrical components.

METER DIFFERENCES: In analog meters the only difference is in the internal circuits of the meter and how the magnetic field is created. All of them require some flow of current through the internal circuits to actuate the meter movement. There must also be an electrical load inside the meter. Each type of meter has some difference for creating or directing the current flow when voltage, amperage, or resistance is being measured. This is caused by making the proper selection with the selector switch, on multimeters. This is not a problem when single use type meters are used. Analog meters have an accuracy of approximately ± 2% of the full scale. Analog meters use a magnetic field to operate the needle movement. The internal circuitry of an analog meter is usually protected from damage by an internal fuse that will blow when an overload is detected. The fuse will need to be replaced before the meter can be used again.

Digital meters are becoming more popular in the air conditioning and refrigeration industry because more solid state controls are being used. The digital meter can be read more accurately than the analog meter. Their cost has also been decreasing over the past years. The smaller voltages used in electronic control systems requires more accuracy in test meters. This type of meter uses a direct read-out on the instrument which does not require that an estimate be made like when the analog type meter is used. See Figure 7-2

***Figure 7-2 Digital (a) and analog (b) meters. (Courtesy of
A.W. Sperry Instruments Inc.)***

Most digital meters use a 3.5 or a 4.5 display. That is to say that there are either four or five digits in the display. The 0.5 is usually blank (0) or 1 which is the left digit of the meter reading. This procedure is used to determine the basic accuracy of the meter. The accuracy of digital meters is usually in the $\pm$ 1% of the actual reading. Digital meters use Ohms law to measure and display the characteristics of the electrical circuit. There are generally two methods used to protect digital meters from overload damage. One is an internal circuit breaker that will detect and overload and open the meter circuits. It will automatically return the meter to operation after the overload has been removed. The other method is to use an internal fuse that will blow and must be changed before the meter can be used again.

Meters are available in single meter cases and multimeter type cases. There are advantages and disadvantages to both types. The more meters in one case the more expensive it is to buy. Also, when one meter needs repair the complete unit must be sent to the repair shop. Thus, making it more difficult to find a temporary replacement. This type meter also requires more time to learn to read it accurately.

When the case contains only one meter it is less expensive to buy. However, when all meters are bought in this way the total expense may be more than that of multiple meters in a single case. The single meters are usually easier to learn to read accurately. When one meter needs repair, only that meter needs to be sent to the repair shop. It is usually easier to find a temporary replacement for a single meter.

The user must make the decision which type meter to buy for personal use.

VOLTMETER: The voltmeter is used to measure the amount of pressure or voltage in a circuit. The air conditioning and refrigeration technician must know what the voltage supplied to a component is so that a proper diagnosis can be made.

There is a wide variety of these instruments available. They range from very small, inexpensive to very complex and very expensive. The amount of accuracy is also a factor in their cost. They are available in a variety of shapes and sizes. The one chosen should meet the needs of the user. Some have only one or two ranges and others have several ranges on one meter scale. The multiple ranges are more complicated to use and read than the smaller units. See Figure 7-3.

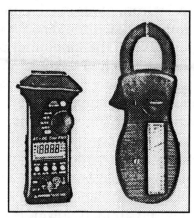

Figure 7-3 Simple and complex voltmeters. (Courtesy of Amprobe Instrument Division of SOS Consolidated Inc.)

Some meters will measure only ac voltage and others will measure both ac and dc voltage. Some are available that will indicate voltage, amperage, and resistance.

Using the Voltmeter: The voltmeter leads are connected directly across the electrical lines for which the voltage is to be measured. See Figure 7-4.

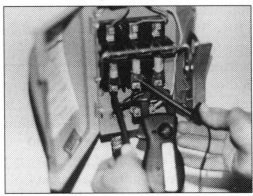

Figure 7-4 Measuring voltage with a voltmeter. (Photo by Billy C. Langley.)

CAUTION: To prevent electrical shock, always use caution when testing any component using the applied voltage. Prevent any bare wire from touching the unit, a component, or personnel. A grounded wire could cause possible damage to the equipment and/or to the personnel present.

Before the leads are touched to the circuit terminals, be sure that the voltage scale selected is capable of measuring the applied voltage without damaging the meter. If the voltage in the system is not known, start with the highest range available and then select the proper scale. The best reading on an analog meter is at the mid-range position. The digital meter will read what voltage is present. However, a more accurate reading will be displayed when the closest range, without going below the applied power, is used. The ranges are provided by the internal circuitry of the meter.

To check fuses and circuit breakers with the voltmeter the power must remain on. Check the voltage from the line terminal to the load side of the fuse. If a voltage is detected the fuse is open and must be replaced. See Figure 7-5.

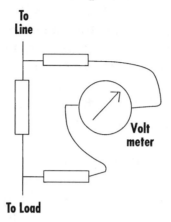

To Line

Volt meter

To Load

Figure 7-5 Checking a fuse with a voltmeter.

As a second test for a fuse and to determine if the line voltage is properly supplied, check from the line side of one fuse to the load side of the fuse. This will indicate if the line voltage is out or if the fuse is blown. When a voltage is indicated the fuse is blown. When no voltage is indicated the fuse is usually good. See Figure 7-6.

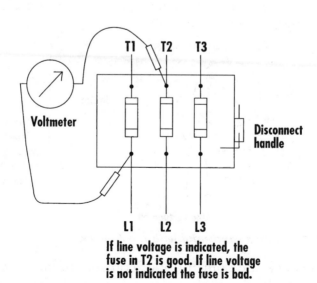

T1 T2 T3

Voltmeter

Disconnect handle

L1   L2   L3

If line voltage is indicated, the fuse in T2 is good. If line voltage is not indicated the fuse is bad.

Figure 7-6 Checking fuses with a voltmeter with the power on.

To check for a voltage drop, connect the meter leads to the terminals of the component being tested and observe the voltage reading while the unit is starting. The voltage may drop slightly during the starting phase and then return to normal. This is normal operation. If the voltage drops below that recommended by the equipment manufacturer during the start phase and does not return to normal immediately after starting, there is a voltage problem that must be corrected.

The voltmeter can also be used to test the voltage across a capacitor. Connect the voltmeter to the terminals of the capacitor and start the unit. After the unit has started read the voltage indicated. This will indicate the minimum voltage requirement of the capacitor.

The voltmeter may also be used to determine if there is continuity through a component, such as an electric motor. To make this test, disconnect one lead from a terminal of the component and connect the meter between the terminal and the removed lead. Apply voltage to the component and if the applied line voltage is indicated there is continuity through the component. This test does not indicate the amount of resistance present but only that there is continuity through it. See Figure 7-7.

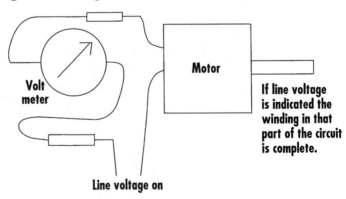

Volt meter

Motor

If line voltage is indicated the winding in that part of the circuit is complete.

Line voltage on

Figure 7-7 Testing continuity with a voltmeter.

To check the continuity of control contacts, connect the meter leads to each contact terminal of the control.

Energize the control circuit. If a voltage is indicated the control is open. Either the contacts are bad or there is some other fault with the contact continuity.

Learning to accurately read the scales on analog meters usually takes time and practice by the user. This will usually come after the meter has been used a while. It is best to practice with the meter until you are comfortable with taking readings with it. This will save time and money when using the meter on the job. The marks on the scale will have different meanings for each different range selected.

AMMETER: The ammeter is used to measure the flow of current through an electric circuit. The ammeter is very important to the air conditioning and refrigeration technician. It can be used as a basic indication of whether or not the motors are operating within their recommended current rating. It is used to detect electrical problems in an operating system. They are available in models that will indicate only amperage, or they can be purchased that will indicate amperage, voltage, and resistance when using the correct leads and/or settings. They are also available in analog and digital reading types.

The basic ammeter movement is the same one used for measuring voltage with the meter. It is the strength of the magnetic field that the current causes as it flows through the meter movement that moves the needle a given amount for each flow of current. The stronger the current flow the stronger the magnetic field produced by it. The stronger magnetic field causes the needle to indicate a higher current flow.

Using the Ammeter: The most popular ammeter is the clamp-on type. See Figure 7-8.

Figure 7-8 Clamp on ammeter. (Courtesy of A.W. Sperry Instruments Inc.)

There are some in-line meters used, but they are usually for specialized purposes. However, the clamp-on type is much easier to use because the wire does not need to be separated to test the current flow. The jaws are simply clamped around the wire feeding the electricity to the device being tested. See Figure 7-9.

Figure 7-9 Checking current flow with a clamp-on ammeter. (Courtesy of Amprobe Instrument Division of SOS Consolidated Inc.)

Clamp-on ammeters are available in both analog and digital types.

To use a clamp-on ammeter, just clamp the jaws of the meter around a single wire to the device being tested. Never place two wires in the jaws because if the current in each wire is flowing in opposite directions, they will tend to cancel out each other and the meter will indicate no current flow. If the two wires are carrying current in the same direction the meter will sense both currents and indicate a higher reading. The jaws of the ammeter sense the magnetic field around the wire carrying the current and transfers this energy to the internal components of the meter. When testing a circuit in which the current flow is not known it is better to set the meter on the highest possible scale and then select the correct scale when a reading is indicated. This will prevent possible damage to the meter from excessive current flow for the range selected. The analog type meter is most accurate when the reading is indicated at midscale.

When a small amperage is being measured it is difficult to read an analog meter accurately. This problem can be helped some by making a coil of the wire being measured and placing the coil around one tong of the meter. See Figure 7-10.

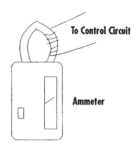

Figure 7-10 Measuring a small current with an ammeter.

Sometimes this can help in getting a more accurate reading when a digital meter is being used. The resulting current reading is then divided by the number of turns in the coil.

Example: We are measuring the current draw in a thermostat circuit. The reading is so small that it cannot be accurately read. Make a coil of ten (10) turns in the wire being tested. Take the reading and find it to be 4.5 amps. To determine the correct amperage, divide the amperage by the number of turns in the coil.

I = 4.5 ÷ 10 = 0.45 amps.

To test the starting and running amperage of a motor, set the ammeter scale on the highest possible scale and place the jaws around the wire to the common terminal of the motor. Start the motor while observing the ammeter. The meter will indicate approximately six times the running amperage for a fraction of a second. Then the meter will indicate the running amperage. Needless to say, the starting amperage must be read very fast.

To check the total unit amperage start the unit and place the ammeter jaws around one of the electric wires going to the unit. Usually this is best accomplished in the terminal compartment of the unit. The amperage indicated will be the total unit running amperage.

To test the amperage of each of the individual components in the unit, determine which component is to be tested first. Then place the ammeter jaws around any one of the wires going to that component. Make certain that no other component is connected into the line between the point where the reading is taken and the component. If there is another connection, the amperage flow to all the components will be indicated. See Figure 7-11.

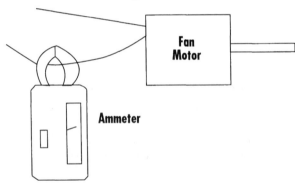

Figure 7-11 Checking amperage on individual components.

OHMMETERS: Ohmmeters are used to determine the amount of resistance a circuit or a component has. See Figure 7-12.

Figure 7-12 A combination meter including an ohmmeter function.
(Courtesy of Amprobe Instrument Division of SOS
Consolidated Inc.)

And to determine if the circuit or component is open or closed. They can also be used to determine if a direct short is present. See Figure 7-13.

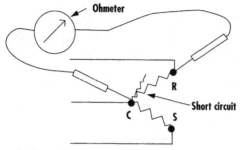

Figure 7-13 Direct short (little or no resistance).

This is sometimes referred to as testing the continuity of the circuit or component. When a circuit or component has continuity it is understood that there is a complete circuit through it. When a reading of infinite resistance is indicated by the meter, the circuit or component being tested does not have a complete path through it. See Figure 7-14.

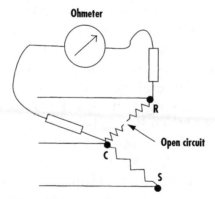

Figure 7-14 Open circuit (Infinite resistance).

When a resistance is indicated, it is usually referred to as a measurable resistance. See Figure 7-15.

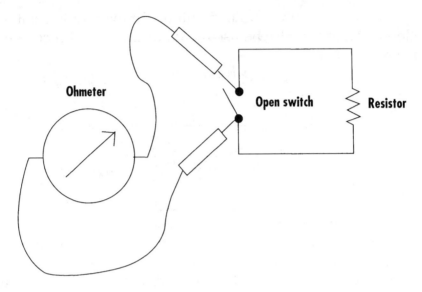

Figure 7-15 Measurable resistance.

Ohmmeters are usually powered by batteries and are designed to operate only on this small current. If they are connected to a source of power other than their battery, damage will usually result. Sometimes it is just a blown fuse and sometimes it is much more serious. Be sure that the power is disconnected from the circuit or component before touching the ohmmeter leads to a terminal. Also, to prevent damage from leaking batteries, be sure to remove them from the meter when it is to be stored.

The meter movement of an ohmmeter is designed to operate on a very low amount of current flow such as that provided by a battery. The ohmmeter scale on the face of the meter is calibrated to indicate the amount of resistance by the amount of current flowing through the meter movement. The amount of power supplied by the battery and the amount of resistance present will determine how far the magnetic field will move the needle. To obtain accurate readings, the ohmmeter must be "zeroed" before each use and after changing scales because the field caused by the battery will be different after each scale change. When the power from the battery decreases the magnetic field used to move the needle will be less. When storing the ohmmeter, make certain that the selector switch is turned to the "off" position, or at least to some point that power will not be used which will run down the battery.

Ohmmeters are available in either single use cases or in multiple use cases. In most cases the ohmmeter, voltmeter, and ammeter are combined into a single case for convenience and use.

Using the ohmmeter: Before using an ohmmeter, disconnect the electrical circuit from the power supply. An ohmmeter can be severely damaged when touched to a voltage other than that supplied by the battery. Also, before using an ohmmeter be sure to zero the meter for the scale to be used. Otherwise, faulty readings will probably be indicated.

The range of the scales on ohmmeters is from 1Ω to $1,000\Omega$. Some special scales are available on special order. The scales on your meter should best serve your uses.

Ohmmeters are used to check for open circuits, shorted circuits, and for resistance in a circuit or a component. The technician must be able to properly use an ohmmeter.

Sometimes it is necessary to use an ohmmeter to locate a shorted circuit because the circuit breaker will be immediately tripped or the fuse will be immediately blown. Not allowing the technician enough time to locate the problem. A shorted circuit occurs when the circuit resistance is removed. Thus, allowing the current to flow directly from one power leg to the other without any resistance. This will cause an overload and possible cause damage to the wiring or the building.

An open circuit only causes a problem in that the equipment will not run or operate properly. There will be no current flowing through the circuit. Open circuits usually occur when a switch has not closed for come reason, a fuse could be blown, or a circuit breaker tripped open. In an open circuit there is no path for the current to flow through.

Many times you will be required to measure the resistance of a component to make certain that it is in good operating condition. To determine the correct resistance rating of motor windings, solenoid coils, and other components, the manufacturers charts for that component must be consulted. The manufacturers usually makes this information available to service technicians.

MEGOHMMETERS: Megohmmeters are available in both battery powered and crank types. The battery powered type is usually recommended. When the crank type is used by an inexperienced person a winding could possibly be ruptured, causing a shorted condition. Their main purpose is to determine if a high resistance short or ground is present. They can also be used to determine if moisture or other contaminants are present in the winding, or wiring. They have been successfully used in scheduled preventative maintenance programs to determine the condition of the refrigeration system as well as the windings of motors and compressor-motors.

Using the Megohmmeter: Make certain that the circuit is disconnected from the power supply.

The Megohmmeter is connected to the system just like a less sensitive meter. It should also be zeroed before use.

The condition of a compressor-motor winding can be tested to determine if contaminants are present in the refrigeration system. The following procedures are recommended for this test:

(1) Read the instrument operating conditions.

(2) Disconnect the power source from the circuit to be tested. Test with a voltmeter to make certain that it has been disconnected.

(3) Disconnect all wires and relays to the compressor terminals.

(4) Connect the black lead of the megohmmeter to the compressor housing.

(5) Touch the red lead to one compressor terminal. Then press the read button, if one is incorporated into the meter.

(6) Hold that position for one minute while observing the reading. A slight increase in the resistance during this time is normal.

(7) Repeat steps 5 and 6 for each terminal on the compressor. Compare these readings to the chart shown in Figure 7-16.

Reading In Megohms	Compressor Condition	Possible Solution
100 – Infinity	Excellent	None
50 – 100	Moisture Present	Change Drier
20 – 50	Moisture Present and Possible Oil Contamination	* Change Drier and Oil
5 – 20	Severe Oil and Moisture Contamination	* Change oil and Driers, Evacuate and recharge system
0 – 5	Possible Insulation Damage (Motor Burnout) Check Carefully	Compressor replacement may be necessary

Note: The table shown above is merely a guide for checking the condition of a sealed system. Readings may vary with certain makes of compressors: slightly above or below the stated limits.

(Some compressor manufacturers publish the insulation resistance specifications.)

* Before changing the oil, an oil analysis is suggested. There are several test kits available for this purpose.

Figure 7-16 *Resistance to motor case insulation indications.*

NOTE: When using a megohmmeter on a scheduled preventative maintenance program, be sure to take readings under similar ambient temperature conditions and for the same amount of time for each test.

CAUTION: Never make this test on a system that is under a vacuum.

Unit 8: Conductors And Insulators

INTRODUCTION: The use of conductors and insulators is what makes it possible to use electric current. If it was not for conductors (wires) the electric current would have no convenient path to follow. The insulation around the conductors is what keeps the electric current flowing in the path for which the circuit was designed.

The atomic structure of each element is different from any other element. The number of neutrons, protons, and electrons and how the electrons are placed in their orbits are different for each element. When the electrons rotating around the nucleus are easily removed from the atom, the element a good conductor. When the outer electrons are not easily removed from the atom, element a good insulator.

CONDUCTORS: Conductors are the materials that provide a good path for the electric current to flow through. They are generally made from materials that have atoms with fewer electrons, usually one or two, in the valence shell. See Figure 8-1.

Copper wire

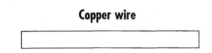

Figure 8-1 Good conductor.

The materials that are considered to make the best conductors are silver, copper, and aluminum. Most metals are considered to be good conductors, however, not all metals will easily conduct electricity. Silver is considered to be the best electrical conductor. However, because of its high cost, silver is seldom used in anything other than for contact surfaces or specialized uses. Copper is considered to be second best for conductor use. It is much less expensive and is used in most electrical conductor applications. There is no perfect conductor. All of them cause some restriction to the flow of electricity through them. The best we can do is design the circuits to prevent as many problems as possible.

INSULATORS: Materials that do not readily transmit electrons are used as insulators. Insulators are used to prevent the flow of electrons into areas where they are not wanted or desired. They are generally made from materials having small, light weight atoms. The atoms of insulating materials will have their valence shells more than half full of electrons. See Figure 8-2.

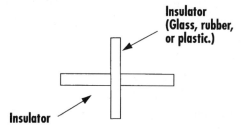

Figure 8-2 Glass, rubber, or plastic are good insulators.

The best insulators will typically have seven or eight electrons in the valence shell. The electrons are held tightly in the orbit and any energy gained is divided between all of them. Making them difficult to remove from the atom.

Materials that make good insulators are glass, rubber, and asbestos. The strength of current flow through a conductor will determine the type of insulator to be used. When the flow of electrons is stronger than what the insulation was designed for, the insulation will break down and allow the current to flow through the wrong path. There is no perfect insulator. All of them will break down under overloaded conditions or where excess heat is present. The best that we can do is to follow recommended guidelines made by authorities.

Unit 9: Wire Size And Voltage Drop

INTRODUCTION: When installing, servicing, or maintaining air conditioning and refrigeration equipment using the proper wire size of great importance. When a wire that is smaller than needed there will be a voltage drop and a resulting drop in the power supplied to the unit. This could possibly cause damage to the unit components. When the wire size is larger than needed the cost will be more expensive than necessary.

WIRE SIZE: In the United States, the standard wire is defined by the American Wire Gauge (AWG). This information is placed in tables that list the wire size, resistance at different temperatures and other data. See Table 9-1.

This table shows that the largest wire is 0000 (4/0) and the smallest wire is a number 50. Seldom does air conditioning and refrigeration applications use smaller than a number 20 wire. These are usually limited to control circuits.

When the application requires a wire size larger than 4/0, the circular mill system is used. This would be on commercial and industrial applications and where a large meter loop is required for the building. The most popular size of wires in this range are from about 250 MCM to 750 MCM (MCM is the abbreviation representing 1000 circular mills). 250 MCM is about 1/2 inch in diameter and 750 MCM is about 1 inch in diameter. This is by no means the only sizes available in the circular mill standard.

Current-Carrying Capacity of a Wire: When discussing the current-carrying capacity of a wire, three factors should be considered: **(1)** the wire resistance will determine the amount of current that will flow through it when connected across a power source.; **(2)** even though the wire resistance will allow a given amount of current to flow, there can also be no more current flowing than the power source can supply. Power sources have a maximum safe limit of current

Table 310-16. Allowable Ampacities of Insulated Conductors Rated 0 through 2000 Volts, 60° to 90°C (140° to 194°F) Not More Than Three Current-Carrying Conductors in Raceway or Cable or Earth (Directly Buried), Based on Ambient Temperature of 30°C (86°F)

Size AWG kcmil	60°C (140°F) COPPER	75°C (167°F) COPPER	90°C (194°F) COPPER	60°C (140°F) ALUMINUM	75°C (167°F) ALUMINUM	90°C (194°F) ALUMINUM	Size AWG kcmil
18	....	....	14	....	....	....	....
16	....	....	18	....	....	....	....
14	20†	20†	25†	....	....	....	....
12	25†	25†	30†	20†	20†	25†	12
10	30	35†	40†	25†	30†	35†	10
8	40	50	55	30	40	45	8
6	55	65	75	40	50	60	6
4	70	85	95	55	65	75	4
3	85	100	110	65	75	85	3
2	95	115	130	75	90	100	2
1	110	130	150	85	100	115	1
1/0	125	150	170	100	120	135	1/0
2/0	145	175	195	115	135	150	2/0
3/0	165	200	225	130	155	175	3/0
4/0	195	230	260	150	180	205	4/0
250	215	255	290	170	205	230	250
300	240	285	320	190	230	255	300
350	260	310	350	210	250	280	350
400	280	335	380	225	270	305	400
500	320	380	430	260	310	350	500
600	355	420	475	285	340	385	600
700	385	460	520	310	375	420	700
750	400	475	535	320	385	435	750
800	410	490	555	330	395	450	800
900	435	520	585	355	425	480	900
1000	455	545	615	375	445	500	1000
1250	495	590	665	405	485	545	1250
1500	520	625	705	435	520	585	1500
1750	545	650	735	455	545	615	1750
2000	560	665	750	470	560	630	2000

CORRECTION FACTORS — For ambient temperatures other than 30°C (86°F), multiply the allowable ampacities shown above by the appropriate factor shown below.

Ambient Temp. °C							Ambient Temp. °F
21-25	1.08	1.05	1.04	1.08	1.05	1.04	70-77
26-30	1.00	1.00	1.00	1.00	1.00	1.00	78-86
31-35	.91	.94	.96	.91	.94	.96	87-95
36-40	.82	.88	.91	.82	.88	.91	96-104
41-45	.71	.82	.87	.71	.82	.87	105-113
46-50	.58	.75	.82	.58	.75	.82	114-122
51-55	.41	.67	.76	.41	.67	.76	123-131
56-60	....	.58	.71	....	.58	.71	132-140
61-70	....	.33	.58	....	.33	.58	141-158
71-80	....	....	.41	....	....	.41	159-176

† Unless otherwise specifically permitted elsewhere in this Code, the overcurrent protection for conductor types marked with an obelisk (†) shall not exceed 15 amperes for No. 14, 20 amperes for No. 12, and 30 amperes for No. 10 copper, or 15 amperes for No. 12 and 25 amperes for No. 10 aluminum and copper-clad aluminum.

Table 310-17. Allowable Ampacities of Single Insulated Conductors, Rated 0 through 2000 Volts, in Free Air Based on Ambient Air Temperature of 30°C (86°F)

Size AWG kcmil	60°C (140°F) COPPER	75°C (167°F) COPPER	90°C (194°F) COPPER	60°C (140°F) ALUMINUM	75°C (167°F) ALUMINUM	90°C (194°F) ALUMINUM	Size AWG kcmil
18	....	....	18	....	....	....	....
16	....	....	24	....	....	....	....
14	25†	30†	35†	....	....	....	....
12	30†	35†	40†	25†	30†	35†	12
10	40†	50†	55†	35†	40†	40†	10
8	60	70	80	45	55	60	8
6	80	95	105	60	75	80	6
4	105	125	140	80	100	110	4
3	120	145	165	95	115	130	3
2	140	170	190	110	135	150	2
1	165	195	220	130	155	175	1
1/0	195	230	260	150	180	205	1/0
2/0	225	265	300	175	210	235	2/0
3/0	260	310	350	200	240	275	3/0
4/0	300	360	405	235	280	315	4/0
250	340	405	455	265	315	355	250
300	375	445	505	290	350	395	300
350	420	505	570	330	395	445	350
400	455	545	615	355	425	480	400
500	515	620	700	405	485	545	500
600	575	690	780	455	540	615	600
700	630	755	855	500	595	675	700
750	655	785	885	515	620	700	750
800	680	815	920	535	645	725	800
900	730	870	985	580	700	785	900
1000	780	935	1055	625	750	845	1000
1250	890	1065	1200	710	855	960	1250
1500	980	1175	1325	795	950	1075	1500
1750	1070	1280	1445	875	1050	1185	1750
2000	1155	1385	1560	960	1150	1335	2000

CORRECTION FACTORS — For ambient temperatures other than 30°C (86°F), multiply the allowable ampacities shown above by the appropriate factor shown below.

Ambient Temp. °C							Ambient Temp. °F
21-25	1.08	1.05	1.04	1.08	1.05	1.04	70-77
26-30	1.00	1.00	1.00	1.00	1.00	1.00	78-86
31-35	.91	.94	.96	.91	.94	.96	87-95
36-40	.82	.88	.91	.82	.88	.91	96-104
41-45	.71	.82	.87	.71	.82	.87	105-113
46-50	.58	.75	.82	.58	.75	.82	114-122
51-55	.41	.67	.76	.41	.67	.76	123-131
56-60	....	.58	.71	....	.58	.71	132-140
61-70	....	.33	.58	....	.33	.58	141-158
71-80	....	....	.41	....	....	.41	159-176

† Unless otherwise specifically permitted elsewhere in this Code, the overcurrent protection for conductor types marked with an obelisk (†) shall not exceed 15 amperes for No. 14, 20 amperes for No.12, and 30 amperes for No. 10 copper, or 15 amperes for No. 12 and 25 amperes for No. 10 aluminum and copper-clad aluminum.

Table 9-1 Allowable ampacities of insulated conductors rated 0 - 2000 volts, 60° to 90°C (National Electrical code® and NEC® are registered trademarks of the National Fire Protection Association, Inc., Quincy, MA 02269) (Reprinted with permission from NFPA 70-1996. The National Electrical code®, copyright© 1995, National Fire Protection Association, Quincy, MA 02269. This reprinted material is not the complete and official position of the National Fire Protection Association, on the referenced subject which is represented only by the Standard in its entirety.

flow that they can supply before burning out. For this reason, a conductor is seldom connected directly across a power source, and **(3)** the amount of current that the conductor can carry must be considered. Remember that an electric current causes a wire to heat. When too much current flows through a wire it becomes hot and the insulation could burn off, ruining the conductor.

It should be remembered that the higher the gauge number, the smaller the cross-sectional area of the wire. Thus, wires with a higher-numbered gauge size offer more resistance the flow of electricity. Table 9-1 shows the diameter corresponding to each gauge number and the resistance per 1000 feet of the wire. The resistance for each wire is given for the normal 77°F (25°C). You should become familiar with the characteristics of the most popular wire sizes used.

Ampacity: Ampacity is defined as the current-carrying capability of a given conductor. The factors that determine the amount of current that a conductor can carry are: the type of material it is made from; its size; and the type of insulation covering it. The National Electrical Code (NEC) governs the minimum ampacities (MCA) for each wire size and its purpose. The MCA ratings can be found in the National Electrical Code manual.

Ampacity ratings are now being placed on the equipment nameplates by the manufacturer. The connecting wires must be sized according to the MCA ratings and the NEC guidelines.

The maximum fuse size (MFS) is also listed on the equipment nameplate. This over current protection should be followed to prevent the needless tripping of circuit breakers and blowing of fuses.

Temperature and Resistance: When determining the resistance of a piece of material its temperature should be given consideration. Most of the metals used in conductors, such as copper and aluminum, have a increase in resistance with an increase in temperature. In many circuits, especially electronic circuits, careful design is required to ensure proper ventilation and radiation of the heat produced by the current-carrying devices. When the conductors are to be enclosed in a conduit, the National Electric Code specifies that larger wire sizes should be used so that heat, due to line losses, will not affect the current-carrying capacity of the conductors.

Trade Name	Type Letter	Maximum Operating Temp.	Application Provisions	Insulation	AWG or kcmil	Thickness of Insulation	Mils	Outer Covering****
Fluorinated Ethylence Propylene	FEP or FEPB	90°C 194°F 200°C 392°F	Dry and damp locations Dry locations—special applications.†	Fluorinated Ethylence Propylene	14-10 8-2		20 30	None
				Fluorinated Ethylence Propylene	14-8 6-2		14 14	Glass Braid Asbestos or other suitable braid material
Mineral Insulation (Metal Sheathed)	MI	90°C 194°F 250°C 482°F	Dry and wet locations For special application.†	Magnesium Oxide	18-16***** 16-10 9-4 3-500		23 36 50 55	Copper or Alloy Steel
Moisture-, Heat-, and Oil-Resistant Thermoplastic	MTW	60°C 140°F 90°C 194°F	Machine tool wiring in wet locations as permitted in NFPA 79. (See Article 670.) Machine tool wiring in dry locations as permitted in NFPA 79. (See Article 670.)	Flame-Retardant, Moisture-, Heat- and Oil Resistant Thermoplastic	22-12 10 8 6 4-2 1-4/0 213-500 501-1000	(A) 30 30 45 60 60 80 95 110	(B) 15 20 30 30 40 50 60 70	(A) None (B) Nylon jacket or equivalent
Paper		85°C 185°F	For underground service conductors, or by special permission.	Paper				Lead Sheath
Perfluoroalkoxy	PFA	90°C 194°F 200°C 392°F	Dry and damp locations. Dry locations—special applications.†	Perfluoroalkoxy	14-10 8-2 1-4/0		20 30 45	None
Perfluoroalkoxy	PFAH	250°C 482°F	Dry locations only. Only for leads within apparatus or within raceways connected to apparatus. (Nickel or nickel-coated copper only.)	Perfluoroalkoxy	14-10 8-2 1-4/0		20 30 45	None
Thermoset	RH	75°C 167°F	Dry and damp locations.	Flame-Retardant Thermoset	14-12** 10 8-2 1-4/0		30 45 60 80	Moisture-resistant, flame-retardant, non-metallic covering.*
Thermoset	RHH	90°C 194°F	Dry and damp locations.		213-500 501-1000 1001-2000 For 601-2000 volts, See Table 310-62		95 110 125	

*Some rubber insulations do not require an outer covering.
**For size Nos. 1-12, RHH shall be 45 mils-thickness insulation.
****Some insulations do not require an outer covering.
*****For signaling circuits permitting 300-volt insulation.
†Where design conditions require maximum conductor operating temperatures above 90°C.

Table 9-1 (Continued).

We can summarize the factors that affect resistance as follows:

1. Length of the conductor
2. Size of the conductor
3. Type of metal
4. Temperature.

Insulation: The type of insulation covering a wire determines to some extent the current-carrying capacity of the wire. This is because some types of insulation can withstand more heat than others. This information is placed in tables for ready reference. See Table 9-2.

Trade Name	Type Letter	Maximum Operating Temp.	Application Provisions	Insulation	AWG or kcmil	Thickness of Insulation Mils	Outer Covering****
Moisture-Resistant Thermoset	RHW †††	75°C 167°F	Dry and wet locations. Where over 2000 volts insulation, shall be ozone-resistant.	Flame Retardant, Moisture-Resistant Thermoset	14-10 8-2 1-4/0 213-500 501-1000 1001-2000 For 601 2000 volts, see Table 310-62	45 60 80 95 110 125	Moisture-Resistant, Flame-Retardant, nonmetallic Covering*
Moisture-Resistant Thermoset	RHW-2	90°C 194°F	Dry and wet locations.	Flame-Retardant, Moisture-Resistant Thermoset	14-10 8-2 1-4/0 213-500 501-1000 1001-2000 For 601 2000 volts, see Table 310-62	45 60 80 95 110 125	Moisture-Resistant, Flame-Retardant, nonmetallic Covering*
Silicone	SA	90°C 194°F 200°C 392°F	Dry and damp locations. For special application.†	Silicone Rubber	14-10 8-2 1-4/0 213-500 501-1000 1001-2000	45 60 80 95 110 125	Glass or other suitable braid material
Thermoset	SIS	90°C 194°F	Switchboard wiring only.	Flame-Retardant, Thermoset	14-10 8-2 6-2 1-4/0	30 45 60 55	None
Thermoplastic and Fibrous outer braid	TBS	90°C 194°F	Switchboard wiring only.	Thermoplastic	14-10 8 6-2 1-4/0	30 45 60 80	Flame-Retardant, nonmetallic Covering
Extruded Polytetra-fluoroethylene	TFE	250°C 482°F	Dry locations only. Only for leads within apparatus or within raceways connected to apparatus, or as open wiring. (Nickel or nickel-coated copper only.)	Extruded Polytetra-fluoroethylene	14-10 8-2 1-4/0	20 30 45	None
Heat-Resistant Thermoplastic	THHN	90°C 194°F	Dry and damp locations.	Flame-Retardant, Heat-Resistant Thermoplastic.	14-12 10 8-6 4-2 1-4/0 250-500 501-1000	15 20 30 40 50 60 70	Nylon jacket or equivalent
Moisture- and Heat-Resistant Thermoplastic	THHW	75°C 167°F 90°C 194°F	Wet locations. Dry locations.	Flame-Retardant, Moisture- and Heat-Resistant Thermoplastic.	14-10 8-2 1-4/0 213-500 501-1000	45 60 80 95 110	None
Moisture- and Heat-Resistant Thermoplastic	THW †††	75°C 167°F 90°C 194°F	Dry and wet locations. Special applications within electric discharge lighting equipment. Limited to 1000 open-circuit volts or less. (Size 14-8 only as permitted in Section 410-31.)	Flame-Retardant, Moisture- and Heat-Resistant Thermoplastic.	14-10 8-2 1-4/0 213-500 501-1000 1001-2000	45 60 80 95 110 125	None
Moisture- and Heat-Resistant Thermoplastic	THWN†††	75°C 167°F	Dry and wet locations.	Flame-Retardant, Moisture- and Heat-resistant Thermoplastic.	14-12 10 8-6 4-2 1-4/0 250-500 501-1000	15 20 30 40 50 60 70	Nylon jacket or equivalent
Moisture-Resistant Thermoplastic	TW	60°C 140°F	Dry and wet locations.	Flame-Retardant, Moisture-Resistant Thermoplastic.	14-10 8 6-2 1-4/0 213-500 501-1000 1001-2000	30 45 60 80 95 110 125	None

*Some rubber insulations do not require an outer covering.
****Some insulations do not require an outer covering.
†Where design conditions require maximum conductor operating temperatures above 90°C.
†††Listed wire types designated with the suffix "-2," such as RHW-2. shall be permitted to be used at a continuous 90°C operating temperature, wet or dry.

Table 9-2 Wire insulating materials (National Electrical code® and NEC® are registered trademarks of the National Fire Protection Association, Inc., Quincy, MA 02269) (Reprinted with permission from NFPA 70-1996. The National Electrical code®, copyright© 1995, National Fire Protection Association, Quincy, MA 02269. This reprinted material is not the complete and official position of the National Fire Protection Association, on the referenced subject which is represented only by the Standard in its entirety.

Trade Name	Type Letter	Maximum Operating Temp.	Application Provisions	Insulation	AWG or kcmil	Thickness of Insulation Mils	Outer Covering****
Underground Feeder and Branch-Circuit Cable Single Conductor. (For Type UF cable employing more than one conductor, see Article 339.)	UF	60°C 140°F	See Article 339.	Moisture-Resistant	14-10 8-2 1-4/0	60* 80* 95*	Integral with insulation
		75°C 167°F**		Moisture- and Heat-Resistant			
Underground Service-Entrance Cable Single Conductor. (For Type USE cable employing more than one conductor, see article 338.)	USE†††	75° C 167°F	See Article 338.	Heat- and Moisture-Resistant	12-10 8-2 1-4/0 213-500 501-1000 1001-2000	45 60 80 95*** 110 125	Moisture-resistant non-metallic covering. [See section 338-1(b).]
Thermoset	XHH	90° C 194°F	Dry and damp locations.	Flame-retardant Thermoset	14-10 8-2 1-4/0 213-500 501-1000 1001-2000	30 45 55 65 80 95	None
Moisture-resistant Thermoset	XHHW†††	90° C 194°F 75°C 167°F	Dry and damp locations. Wet locations.	Flame-retardant Moisture-Resistant Thermoset	14-10 8-2 1-4/0 213-500 501-1000 1001-2000	30 45 55 65 80 95	None
Moisture-resistant Thermoset	XHHW-2	90° C 194°F	Dry and wet locations.	Flame-retardant Moisture-Resistant Thermoset	14-10 8-2 1-4/0 213-500 501-1000 1001-2000	30 45 55 65 80 95	None
Modified Ethylene Tetrafluoroethylene	Z	90° C 194°F 150°C 302°C	Dry and damp locations. Dry locations— special applications.†	Modified Ethylene Tetrafluoroethylene	14-12 10 8-4 3-1 1/0-4/0	15 20 25 35 45	None
Modified Ethylene Tetrafluoroethylene	ZW†††	75°C 167°F 90°C 194°F 150°C 302°F	Wet locations. Dry and damp locations. Dry locations—special applications.†	Modified Ethylene Tetrafluoroethylene	14-10 8-2	30 45	None

*Includes integral jacket.
**For ampacity limitation, see Section 339-5.
***Insulation thickness shall be permitted to be 80 mils for listed Type USE conductors that have been subjected to special investigations.
The nonmetallic covering over individual rubber-covered conductors of aluminum-sheathed cable and of lead-sheathed or multiconductor cable shall not be required to the flame-retardant. For Type MC cable, see Section 334-20. For nonmetallic-sheathed cable, see Section 336-25. For type UF cable, see Section 339-1.
†Where design conditions require maximum conductor operating temperatures above 90°C.
†††Listed wire types designated with the suffix "-2," such as RHW-2, shall be permitted to be used at a continuous 90°C operating temperature, wet or dry.

Table 9-2 (continued).

In this table the type of insulation is listed at the top of each column. Each type of insulation has different temperature and amperage ratings. Number 8 wire is very popular in air conditioning and refrigeration work. The characteristics for each different type of insulation can be determined by reading across the table from the extreme left hand side. Notice how the type of insulation affects the current-carrying capacity of the wire.

Voltage Rating: The voltage rating of a wire is determined by the type of insulation used on it. The type or size of the wire is not considered when determining its voltage rating. Almost all of the wire used in air conditioning and refrigeration work has a voltage rating of 600 volts. This is the amount of voltage that the insulation material can effectively contain. This is generally determined by the type of material the insulation is made from and its thickness.

VOLTAGE DROP: The voltage drop in an electrical circuit is very important. It can cause many problems from motors overheating to components burning out. Any voltage drop between the power supply and the equipment represents a loss in power to the equipment and must be avoided, even small drops. As the voltage goes down the current draw goes up. When the current goes up excessive heat is generated in the equipment.

Testing for Voltage Drop: The voltmeter is used to test for voltage drop. First turn the equipment off and measure the voltage at the electrical panel. See Figure 9-1.

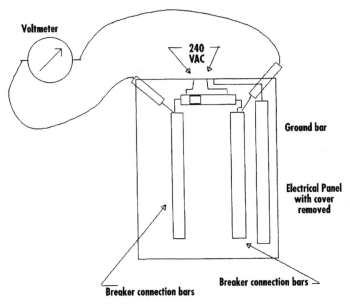

Figure 9-1 Measuring voltage at the electrical panel.

The voltage should be close to what the equipment manufacturer recommends. If it is satisfactory, it can be assumed that there is no problem with the supply. Next, measure the voltage in the panel while the unit is starting. This is best because this is the time of greatest current draw. If the voltage remains satisfactory, it can be assumed that there are no loose or bad connections in that part of the circuit. Should the voltage drop below the recommended voltage, it is assumed that there are bad connections, or that the entrance service is too small for the equipment load. Next, repeat the above steps at the outdoor unit. See Figure 9-1. If the voltage drops below the minimum recommended by the manufacturer indicates that the wire size is too small and that too much voltage is being used to push the current through the wire. This problem can be eliminated by replacing the wires from the electric panel to the outdoor unit.

Unit 10: Electrical Circuits

INTRODUCTION: For electricity to be of use to us, it must be caused to do something that will help us in some manner. In order for electricity to do this it must be directed, controlled, and converted to other forms of energy. The method used to cause this change is the electric circuit. The circuit directs the current to the desired place, the circuit components control the flow, and the motors, heat strips, and other loads, convert the electricity to some form of energy that we can use.

THE ELECTRIC CIRCUIT: The basic electric circuit is made up of three devices: **(1)** the load, **(2)** the conductors, and **(3)** the source of power. The *load* is some device that changes the electrical energy into some other form of energy that is useful to us. The *conductors* are the wires, or path, that carry the electricity from the source to the load. The *power source* supplies the electricity to the circuit where it can be used. See Figure 10-1.

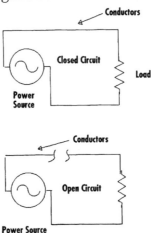

Figure 10-1 The basic electric circuit.

For the current to flow through the circuit there must be a complete path from the negative terminal of the power source, through the conductors and the load, and back to the positive terminal of the power source. If this path is not complete, the circuit is said to be open, and no current will flow through it.

THE SWITCH: The purpose of a switch in an electric circuit is to complete the circuit when electricity is demanded and to open the circuit when the control device is satisfied. The switch is not a load and because of this it uses no electricity. A switch is made up of a set of contacts that open and close the electric circuit. See Figure 10-2.

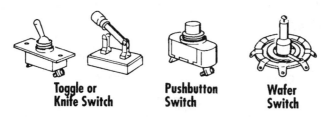

Toggle or Knife Switch   **Pushbutton Switch**   **Wafer Switch**

Figure 10-2 Types of electric switches.

There are many different types of switches available depending on the requirements of the system. The schematic symbols for each of these types is shown in Figure 10-3.

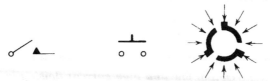

Figure 10-3 Schematic symbols for some switches.

THE CIRCUIT LOAD: The load in an electrical circuit is the device that changes the electrical energy into some type of energy that we can use. An electric motor changes the electrical energy into mechanical energy; light for us to see by; and heat as in an electric furnace or heating element. The electricity may also be used to change or control the amount of electrical output by the power source.

The type of load used will determine the amount of electricity used. In air conditioning and refrigeration, all electric motors use electricity. However, the compressor motor uses more than the fan motor. The heat strips in an electric heater are also heavy users of electricity.

CONTROLLING CURRENT FLOW: In most cases, electrical circuits are designed with a specific purpose in mind and therefore a definite amount of current is required. When an incorrect amount of current flows through the circuit, the circuit will not function as designed. When too much current flows through the load device, the motors, heating elements, etc., or the voltage source could be damaged. When too little current flows through the circuit, the load will not function as designed or perhaps not at all. See Figure 10-4.

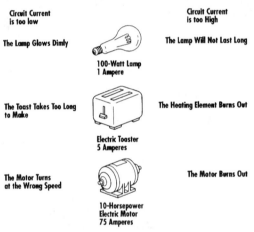

Circuit Current is too low		Circuit Current is too High
The Lamp Glows Dimly	100-Watt Lamp 1 Ampere	The Lamp Will Not Last Long
The Toast Takes Too Long to Make	Electric Toaster 5 Amperes	The Heating Element Burns Out
The Motor Turns at the Wrong Speed	10-Horsepower Electric Motor 75 Amperes	The Motor Burns Out

The current flowing in any circuit must be controlled if the circuit is to work properly

Figure 10-4 Comparison of low and high current flowing in a circuit.

In a DC circuit there are only two things that control the flow of current through the circuit: **(1)** the amount of current supplied by the power source, and **(2)** the conductivity of the wires and the load.

THE SOURCE OF POWER: The power source supplies electricity to the circuit to produce the desired effects. The electricity may be produced by chemical, magnetic, or any of the methods discussed earlier. The energy supplied is generally measured in the difference of electrical potential between the output terminals of the power source. This difference is termed *electromotive force (EMF)* and is measured in volts. The direction of current flow through the circuit is determined by the polarity of the source. Also, the amount of current flowing through the circuit is determined by the amount of voltage being supplied by the power source.

CONDUCTANCE: In our study of electricity, we are mainly concerned of two types of material, conductors and insulators. Remember from our earlier study of electrical theory that not all materials conduct electricity in the same quantity. Conductors are those materials that allow electric current to flow with little resistance, and insulators oppose the flow of electrical current.

The number of free electrons in a material determines its conductivity. Thus, most metals are good conductors. Some metals have more free electrons than others and are better conductors. Conductance is the term used to describe how easy electric current can flow through a material. A high conductance rating means that more current can flow through that material than a material with a low conductance rating. Of the commercially available metals. Silver has the greatest conductance rating. However, copper is more often used because it costs less. See Table 10-1.

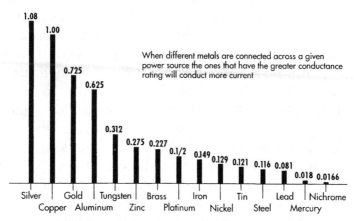

Table 10-1 The relative conductances of different materials.

Silver is usually used on control contacts to reduce the resistance through the circuit and through the control.

Unit 11: Sizing Wire For An Electric Circuit

Most equipment manufacturers usually provide information on the correct wire and fuse size for the unit being installed. However, the installation technician must determine the maximum length of wire required. When the correct wire and fuse size is used the equipment will usually operate trouble-free over its designed left span.

The most popular wire used in the air conditioning and refrigeration industry is copper. Aluminum is sometimes used because of its lower cost. However, when aluminum is used the wire size must be increased according to the NEC. Copper wire is popular because it is pliable and easy to work with, it is strong, it resists corrosion, the conductors can be easily fastened together, and is relatively less expensive than some of the other materials that may be used for conducting electricity. Aluminum does not have these desirable characteristics. It is a good conductor of electricity, but it corrodes very easily. Any connections made involving aluminum wiring will usually loosen after they have become corroded or because of the heat caused by the resistance of the connection. This heat occurs in any type of material because of the resistance to current flow through the junction.

WIRING FACTORS: The NFPA (National Fire Protection Agency) determines what type of wire, with what type of insulation is to be used in any given situation. When in doubt always use their recommendations when wiring a circuit for an air conditioning or refrigeration system.

INSULATION: In most cases it is the insulation that determines the application of the wire. There are several different grades of insulation available which will determine its use. As an example, heat resistant, moisture resistant, heat and moisture resistant, or oil resistant. See Table 11-1.

Table 310-16. Allowable Ampacities of Insulated Conductors Rated 0 through 2000 Volts, 60°C (140°F) to 90°C (194°F) Not More Than Three Current-Carrying Conductors in Raceway or Cable or Earth (Directly Buried), Based on Ambient Temperature of 30°C (86°F)

Size AWG kcmil	COPPER			ALUMINUM OR COPPER-CLAD ALUMINUM			Size AWG kcmil
	60°C (140°F) TYPES TW†, UF†	75°C (167°F) TYPES FEPW†, RH†, RHW†, THHW†, THW†, THWN†, XHHW†, USE†, ZW†	90°C (194°F) TYPES TBS, SA, SIS, FEP†, FEPB†, MI, RHH†, RHW-2, THHN†, THHW†, THW-2†, THWN-2†, USE-2, XHH, XHHW†, XHHW-2, ZW-2	60°C (140°F) TYPES TW†, UF†	75°C (167°F) TYPES RH†, RHW†, THHW†, THW†, THWN†, XHHW†, USE†	90°C (194°F) TYPES TBS, SA, SIS, THHN†, THHW†, THW-2, THWN-2, RHH†, RHW-2, USE-2, XHH, XHHW, XHHW-2, ZW-2	
18	....	....	14	....	....	....	
16	....	....	18	....	....	....	
14	20†	20†	25†	....	....	....	
12	25†	25†	30†	20†	20†	25†	12
10	30	35†	40†	25†	30†	35†	10
8	40	50	55	30	40	45	8
6	55	65	75	40	50	60	6
4	70	85	95	55	65	75	4
3	85	100	110	65	75	85	3
2	95	115	130	75	90	100	2
1	110	130	150	85	100	115	1
1/0	125	150	170	100	120	135	1/0
2/0	145	175	195	115	135	150	2/0
3/0	165	200	225	130	155	175	3/0
4/0	195	230	260	150	180	205	4/0
250	215	255	290	170	205	230	250
300	240	285	320	190	230	255	300
350	260	310	350	210	250	280	350
400	280	335	380	225	270	305	400
500	320	380	430	260	310	350	500
600	355	420	475	285	340	385	600
700	385	460	520	310	375	420	700
750	400	475	535	320	385	435	750
800	410	490	555	330	395	450	800
900	435	520	585	355	425	480	900
1000	455	545	615	375	445	500	1000
1250	495	590	665	405	485	545	1250
1500	520	625	705	435	520	585	1500
1750	545	650	735	455	545	615	1750
2000	560	665	750	470	560	630	2000

CORRECTION FACTORS

Ambient Temp. °C	For ambient temperatures other than 30°C (86°F), multiply the allowable ampacities shown above by the appropriate factor shown below.						Ambient Temp. °F
21-25	1.08	1.05	1.04	1.08	1.05	1.04	70-77
26-30	1.00	1.00	1.00	1.00	1.00	1.00	78-86
31-35	.91	.94	.96	.91	.94	.96	87-95
36-40	.82	.88	.91	.82	.88	.91	96-104
41-45	.71	.82	.87	.71	.82	.87	105-113
46-50	.58	.75	.82	.58	.75	.82	114-122
51-55	.41	.67	.76	.41	.67	.76	123-131
56-60	....	.58	.71	....	.58	.71	132-140
61-70	....	.33	.58	....	.33	.58	141-158
71-80	....	....	.41	....	....	.41	159-176

† Unless otherwise specifically permitted elsewhere in this Code, the overcurrent protection for conductor types marked with an obelisk (†) shall not exceed 15 amperes for No. 14, 20 amperes for No.12, and 30 amperes for No. 10 copper, or 15 amperes for No. 12 and 25 amperes for No. 10 aluminum and copper-clad aluminum.

Table 310-17. Allowable Ampacities of Single Insulated Conductors, Rated 0 through 2000 Volts, in Free Air Based on Ambient Air Temperature of 30°C (86°F)

Size AWG kcmil	COPPER			ALUMINUM OR COPPER-CLAD ALUMINUM			Size AWG kcmil
	60°C (140°F) TYPES TW†, UF†	75°C (167°F) TYPES FEPW†, RH†, RHW†, THHW†, THW†, THWN†, XHHW†, ZW†	90°C (194°F) TYPES TBS, SA, SIS, FEP†, FEPB†, MI, RHH†, RHW-2, THHN†, THHW†, THW-2, THWN-2, USE-2, XHH, XHHW†, XHHW-2, ZW-2	60°C (140°F) TYPES TW†, UF†	75°C (167°F) TYPES RH†, RHW†, THHW†, THW†, THWN†, XHHW†	90°C (194°F) TYPES TBS, SA, SIS, THHN†, THHW†, THW-2, THWN-2, RHH†, RHW-2, USE-2, XHH, XHHW†, ZW-2	
18	....	....	18	....	....	....	
16	....	....	24	....	....	....	
14	25†	30†	35†	....	....	....	
12	30†	35†	40†	25†	30†	35†	12
10	40†	50†	55†	35†	40†	40†	10
8	60	70	80	45	55	60	8
6	80	95	105	60	75	80	6
4	105	125	140	80	100	110	4
3	120	145	165	95	115	130	3
2	140	170	190	110	135	150	2
1	165	195	220	130	155	175	1
1/0	195	230	260	150	180	205	1/0
2/0	225	265	300	175	210	235	2/0
3/0	260	310	350	200	240	275	3/0
4/0	300	360	405	235	280	315	4/0
250	340	405	455	265	315	355	250
300	375	445	505	290	350	395	300
350	420	505	570	330	395	445	350
400	455	545	615	355	425	480	400
500	515	620	700	405	485	545	500
600	575	690	780	455	540	615	600
700	630	755	855	500	595	675	700
750	655	785	885	515	620	700	750
800	680	815	920	535	645	725	800
900	730	870	985	580	700	785	900
1000	780	935	1055	625	750	845	1000
1250	890	1065	1200	710	855	960	1250
1500	980	1175	1325	795	950	1075	1500
1750	1070	1280	1445	875	1050	1185	1750
2000	1155	1385	1560	960	1150	1335	2000

CORRECTION FACTORS

Ambient Temp. °C	For ambient temperatures other than 30°C (86°F), multiply the allowable ampacities shown above by the appropriate factor shown below.						Ambient Temp. °F
21-25	1.08	1.05	1.04	1.08	1.05	1.04	70-77
26-30	1.00	1.00	1.00	1.00	1.00	1.00	78-86
31-35	.91	.94	.96	.91	.94	.96	87-95
36-40	.82	.88	.91	.82	.88	.91	96-104
41-45	.71	.82	.87	.71	.82	.87	105-113
46-50	.58	.75	.82	.58	.75	.82	114-122
51-55	.41	.67	.76	.41	.67	.76	123-131
56-60	....	.58	.71	....	.58	.71	132-140
61-70	....	.33	.58	....	.33	.58	141-158
71-80	....	....	.41	....	....	.41	159-176

† Unless otherwise specifically permitted elsewhere in this Code, the overcurrent protection for conductor types marked with an obelisk (†) shall not exceed 15 amperes for No. 14, 20 amperes for No.12, and 30 amperes for No. 10 copper, or 15 amperes for No. 12 and 25 amperes for No. 10 aluminum and copper-clad aluminum.

Table 11-1 American Wire Gauge (A.W.G.) Working Table (U.S. Bureau of Standards) (National Electrical code® and NEC® are registered trademarks of the National Fire Protection Association, Inc., Quincy, MA 02269) (Reprinted with permission from NFPA 70-1996. The National Electrical code®, copyright© 1995, National Fire Protection Association, Quincy, MA 02269.

This is a partial table from the National Electrical Code (NEC) which lists conductors and their insulating material and their appropriate use.

There are several factors that must be considered when sizing the wire for a given circuit. Among these factors are the voltage drop that will occur in the circuit; the type of insulation on the wire; how it is enclosed, what type of conduit is used; and of course safety. The voltage drop of the total circuit which involves the length of the wires must be determined by the installing personnel. The remainder of the factors are available in tables produced by the National Electrical Code. Wire sizing tables are also published in the National Electrical Code book which include the maximum Ampacity (amperage) for both copper and aluminum conductors.

SIZING THE WIRE: The standard wire size used in the United States is defined by the American Wire Gauge (A.W.G.). Their tables list wire sizes starting with the largest size 0000 (usually pronounced as 4/0), to the smallest size, number 50. The most popular wire sizes used in the air conditioning and refrigeration industry range from about number 20 to number 4/0. In some installations larger wire is needed to properly carry the amperage required by the circuit.

When the larger sizes are needed the circular mil system is used to indicate the wire size. The sizes that include the circular mil measurements used for air conditioning and refrigeration systems range from 250 MCM (MCM is an abbreviation used to indicate 1000 circular mils), to about 750 MCM. 250 MCM wire is about 1/2 inch in diameter. 750 MCM wire is about 1 inch in diameter. There are larger sizes available, but are seldom used in air conditioning and refrigeration systems. See Table 11-2.

VOLTAGE DROP: The voltage drop in a circuit is of major concern when sizing the wire. Any voltage drop results in lost energy through heat. It provides no energy to the equipment. When a large voltage drop occurs, the operation of the equipment will be seriously affected. Even small voltage drops must be avoided. Usually the accepted voltage drop or rise is about 10% above or below the rating of the electricity supplied. The voltage drop can be determined by measuring the voltage at the power source (circuit panel at the meter loop) when the unit is not running. Start the unit and again measure the voltage at the source. Subtract the running voltage from the idle system voltage. When the voltage drop is within the recommended minimum/maximum range is an indication that there is sufficient voltage at the panel and that the connections and wiring for that part of the circuit are good. An indication of excessive voltage drop at the source indicates that there is trouble in that part of the circuit that must be repaired. A voltage drop in excess of 10% of the applied voltage is considered to be the maximum allowed.

Next, check the voltage at the unit terminals with the unit off. Start the unit and again measure the voltage at the unit terminals. Subtract the running voltage from the idle voltage. A considerable voltage drop at this point indicates that the wiring between the power source and the unit is too small. Too much voltage is being used to push the current through the wire. To correct the problem the wire must be replaced with the proper size.

WIRE SIZING CHARTS: An example of a wire sizing chart is shown in Table 11-2.

Fluorinated Ethylene Propylene	FEP or FEPB	90°C 194°F 200°C 392°F	Dry and damp locations Dry locations—special applications.†	Fluorinated Ethylene Propylene	14-10 8-2	20 30	None
				Fluorinated Ethylene Propylene	14-8 6-2	14 14	Glass Braid Asbestos or other suitable braid material
Mineral Insulation (Metal Sheathed)	MI	90°C 194°F 250°C 482°F	Dry and wet locations For special application.†	Magnesium Oxide	18-16***** 16-10 9-4 3-500	23 36 50 55	Copper or Alloy Steel
Moisture-, Heat-, and Oil-Resistant Thermoplastic	MTW	60°C 140°F 90°C 194°F	Machine tool wiring in wet locations as permitted in NFPA 79. (See Article 670.) Machine tool wiring in dry locations as permitted in NFPA 79. (See Article 670.)	Flame-Retardant, Moisture-, Heat- and Oil Resistant Thermoplastic	22-12 10 8 6 4-2 1-4/0 213-500 501-1000	(A) 30 30 45 60 60 80 95 110	(B) 15 20 30 30 40 50 60 70	(A) None (B) Nylon jacket or equivalent
Paper		85°C 185°F	For underground service conductors, or by special permission.	Paper			Lead Sheath
Perfluoroalkoxy	PFA	90°C 194°F 200°C 392°F	Dry and damp locations. Dry locations—special applications.†	Perfluoroalkoxy	14-10 8-2 1-4/0	20 30 45	None
Perfluoroalkoxy	PFAH	250°C 482°F	Dry locations only. Only for leads within apparatus or within raceways connected to apparatus. (Nickel or nickel-coated copper only.)	Perfluoroalkoxy	14-10 8-2 1-4/0	20 30 45	None
Thermoset Thermoset	RH RHH	75°C 167°F 90°C 194°F	Dry and damp locations. Dry and damp locations.	Flame-Retardant Thermoset	14-12** 10 8-2 1-4/0 213-500 501-1000 1001-2000 For 601-2000 volts, See Table 310-62	30 45 60 80 95 110 125	Moisture-resistant, flame-retardant, non-metallic covering.*

*Some rubber insulations do not require an outer covering.
**For size Nos. 1-12, RHH shall be 45 mils-thickness insulation.
****Some insulations do not require an outer covering.
*****For signaling circuits permitting 300-volt insulation.
†Where design conditions require maximum conductor operating temperatures above 90°C.

Table 11-2 Wire Sizing Chart.

Trade Name	Type Letter	Maximum Operating Temp.	Application Provisions	Insulation	AWG or kcmil	Thickness of Insulation Mils	Outer Covering****
Moisture-Resistant Thermoset	RHW †††	75°C 167°F	Dry and wet locations. Where over 2000 volts insulation, shall be ozone-resistant.	Flame Retardant, Moisture-Resistant Thermoset	14-10 8-2 1-4/0 213-500 501-1000 1001-2000 For 601 2000 volts, see Table 310-62	45 60 80 95 110 125	Moisture-Resistant, Flame-Retardant, nonmetallic Covering*
Moisture-Resistant Thermoset	RHW-2	90°C 194°F	Dry and wet locations.	Flame-Retardant, Moisture-Resistant Thermoset	14-10 8-2 1-4/0 213-500 501-1000 1001-2000 For 601 2000 volts, see Table 310-62	45 60 80 95 110 125	Moisture-Resistant, Flame-Retardant, nonmetallic Covering*
Silicone	SA	90°C 194°F 200°C 392°F	Dry and damp locations. For special application.†	Silicone Rubber	14-10 8-2 1-4/0 213-500 501-1000 1001-2000	45 60 80 95 110 125	Glass or other suitable braid material
Thermoset	SIS	90°C 194°F	Switchboard wiring only.	Flame-Retardant, Thermoset	14-10 8-2 6-2 1-4/0	30 45 60 55	None
Thermoplastic and Fibrous outer braid	TBS	90°C 194°F	Switchboard wiring only.	Thermoplastic	14-10 8 6-2 1-4/0	30 45 60 80	Flame-Retardant, nonmetallic Covering
Extruded Polytetra-fluoroethylene	TFE	250°C 482°F	Dry locations only. Only for leads within apparatus or within raceways connected to apparatus, or as open wiring. (Nickel or nickel-coated copper only.)	Extruded Polytetra-fluoroethylene	14-10 8-2 1-4/0	20 30 45	None
Heat-Resistant Thermoplastic	THHN	90°C 194°F	Dry and damp locations.	Flame-Retardant, Heat-Resistant Thermoplastic.	14-12 10 8-6 4-2 1-4/0 250-500 501-1000	15 20 30 40 50 60 70	Nylon jacket or equivalent
Moisture- and Heat-Resistant Thermoplastic	THHW	75°C 167°F 90°C 194°F	Wet locations. Dry locations.	Flame-Retardant, Moisture- and Heat-Resistant Thermoplastic.	14-10 8-2 1-4/0 213-500 501-1000	45 60 80 95 110	None
Moisture- and Heat-Resistant Thermoplastic	THW †††	75°C 167°F 90°C 194°F	Dry and wet locations. Special applications within electric discharge lighting equipment. Limited to 1000 open-circuit volts or less. (Size 14-8 only as permitted in Section 410-31.)	Flame-Retardant, Moisture- and Heat-Resistant Thermoplastic.	14-10 8-2 1-4/0 213-500 501-1000 1001-2000	45 60 80 95 110 125	None
Moisture- and Heat-Resistant Thermoplastic	THWN†††	75°C 167°F	Dry and wet locations.	Flame-Retardant, Moisture- and Heat-resistant Thermoplastic.	14-12 10 8-6 4-2 1-4/0 250-500 501-1000	15 20 30 40 50 60 70	Nylon jacket or equivalent
Moisture-Resistant Thermoplastic	TW	60°C 140°F	Dry and wet locations.	Flame-Retardant, Moisture-Resistant Thermoplastic.	14-10 8 6-2 1-4/0 213-500 501-1000 1001-2000	30 45 60 80 95 110 125	None

*Some rubber insulations do not require an outer covering.
****Some insulations do not require an outer covering.
†Where design conditions require maximum conductor operating temperatures above 90°C.
†††Listed wire types designated with the suffix "-2," such as RHW-2, shall be permitted to be used at a continuous 90°C operating temperature, wet or dry.

Table 11-2 (continued).

This table lists the ampacities (amperages) of a given sized conductor which is made from either copper or aluminum. The NEC wire sizing charts are accurate for most air conditioning and refrigeration systems unless the wire is excessively long. As an example, a three ton condensing unit is rated for 15 amps, the installation technician should add another 25% to the full load amperage rating as a safety factor. This would indicate that a circuit rated for 18.75 amps is required. From Table 11-2, in the column for 60°C with type TW insulation the wire size required is less than the 20 amps listed. Therefore, we can use a number 14 A.W.G. copper wire or a number 12 A.W.G. aluminum wire.

Manufacturers always include the electrical data both in the installation instructions and on the unit electrical panel cover. See Table 11-3.

ELECTRICAL DATA

Model No.		10AC12	10AC18	10AC24	10AC30
Line voltage data		208/230v 60hz-1ph	208/230v 60hz-1ph	208/230v 60hz-1ph	208/230v 60hz-1ph
Compressor	Rated load amps	5.0	8.6	9.8	12.2
	Power factor	.97	.97	.96	.99
	Locked rotor amps	26.3	49.0	56.0	71.0
Condenser Coil Fan Motor	Full load amps	1.1	1.1	1.1	1.1
	Locked rotor amps	1.7	1.7	1.7	1.7
Rec. max. fuse or circuit breaker size (amps)		15	20	20	25
*Minimum circuit ampacity		7.4	12.0	13.4	16.4

*Refer to National Electrical Code manual to determine wire, fuse and disconnect size requirements.
NOTE — Extremes of operating range are plus 10% and minus 5% of line voltage.

ELECTRICAL DATA

Model No.		10AC36	10AC42	10AC48	10AC60
Line voltage data — 60 hz		208/230v 60hz-1ph	208/230v 1ph	208/230v 1ph	208/230v 1ph
Compressor	Rated load amps	16.3	22.0	22.5	30.8
	Power factor	.99	.99	.97	.98
	Locked rotor amps	86.7	105.0	110.0	147.0
Condenser Coil Fan Motor	Full load amps	1.1	1.1	1.7	1.7
	Locked rotor amps	1.7	1.7	3.1	3.1
Rec. max. fuse or circuit breaker size (amps)		35	50	50	60
*Minimum circuit ampacity		21.5	28.6	30.0	40.2

*Refer to National Electrical Code manual to determine wire, fuse and disconnect size requirements.
NOTE — Extremes of operating range are plus 10% and minus 5% of line voltage.

INSTALLATION CLEARANCES (inches)

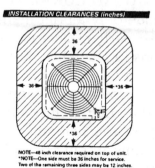

NOTE—48 inch clearance required on top of unit.
*NOTE—One side must be 36 inches for service.
Two of the remaining three sides may be 12 inches.

Table 11-3 Electrical data. (Courtesy of Lennox Industries, Inc.)

The electrical data usually includes the minimum circuit ampacity, the recommended maximum fuse or circuit breaker size (in amps) and the rated load amps for the compressor and fan, along with the locked rotor amps for both. The installation technician will find this information valuable for a proper installation.

Unit 12: Alternating Current

There are two types of electricity used in modern air conditioning and refrigeration systems; they are alternating current (ac), and direct current (dc). The alternating current is used to operate most electric motors. Direct current is used mostly in control circuits. Some manufacturers are now using dc to power compressor motors. However, it is not in wide spread use at this time.

Alternating current (ac) means that the current flows in one direction for a period of time and then reverses its direction for an equal amount of time. In a dc circuit, the current flows in only one direction.

The electricity produced in the United States is alternating 60 cycle (Hz). That means that the current changes its direction of flow 60 times per second.

Alternating current is so popular in the air conditioning and refrigeration that the personnel employed in this industry have a thorough working knowledge of the theory upon which it operates. They should be familiar with the distribution of electricity by the power companies and with the different voltages that are available.

ALTERNATING CURRENT THEORY: Up to this point we have made only a brief mention of alternating current. As described earlier, electric current is the movement of free electrons in a conductor with each end connected to a different potential.

In an electric circuit, as long as there is a potential difference in the circuit, current will flow through it. Should the polarity of the potential remain constant the current will always flow in only one direction and is called direct current (dc).

However, when we consider alternating current remember that the current does not always flow in a single direction. It periodically reverses direction. It flows in one direction for a given period of time and then it reverses and flows in the other. See Figure 12-1.

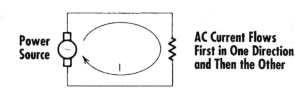 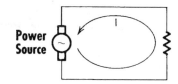

Figure 12-1 AC current.

Also, remember that in an electric circuit the current flows from the negative power source terminal to the positive terminal. From this it can be seen that when alternating current is used, that for the current to flow the polarity of the power source must also change polarity to cause the current to alternate, or change directions.

It may seem that reversing the flow of electrons would cancel any work they had done before being reversed. However, the flowing electrons themselves do not do any useful work. It is the effects that they have on the circuit loads as they flow through them. The effects caused by the flowing electrons are the same regardless of the direction in which they are flowing. It must be remembered that heat is always produced when and electric current passes through a resistance of any type. Thus, the direction of flow of the electrons makes no difference on the amount of heat they produce, or the amount of work done, when flowing through the circuit.

COMPARISON OF AC AND DC: When comparing ac to dc we find that dc current flows in only one direction through the circuit and that ac current periodically reverses its direction of flow through the circuit.

In a dc circuit the current never changes direction nor does its polarity change. See Figure 12-2.

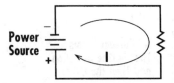

Figure 12-2 DC current.

What is meant by dc current never changes direction is that the voltage starts out a 0 voltage increases either to the maximum negative or maximum positive value and remains at that value. See Figure 12-3.

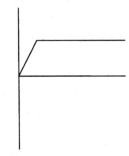

Figure 12-3 DC current graph.

When the voltage is turned off, the value returns to the 0 position. It remains there until the power is again turned on.

The ac power source periodically changes direction. That is when the power is turned on the current starts at 0 and builds to a maximum positive value, returns to 0 and builds to a maximum negative value and then returns to the 0 position for another cycle. This cycle will continue until the power is turned off. See Figure 12-4.

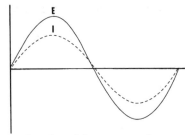

Figure 12-4 Graph showing AC current alternations.

ALTERNATING CURRENT USES: About 90% of all electricity used in the world is ac. There are basically three reasons for this:(1) ac can do almost everything that dc can do, (2) ac can be transferred from the generating plant to the point of use much more economically and easier than dc, and (3) ac can perform certain functions that dc cannot perform.

Direct current electricity will always be used in certain types of application. It is used widely inside electrical equipment to perform functions which are accomplished only with dc power.

DELTA AND WYE CONNECTIONS: These are two methods of connecting the windings of alternators(ac generators). Each type of connection provides different power outputs. *The delta connection* is used where high current requirements are to be met and the *WYE connection* is used to provide high-voltage output.

When the delta system is used, the output voltage equals the winding voltage. See Figure 12-5.

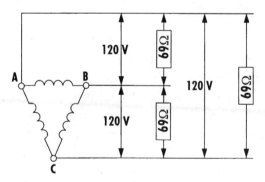

Figure 12-5 Delta system connections.

Notice that there are 120 V between each of the three phase connections. In a delta system the current flow is divided. When this type of connection is used, the alternator winding current equals the line current divided by the square root of three √ 3:

a = ft ÷ √3

The output windings of an alternator could also be

connected to a wye connection. See Figure 12-6.

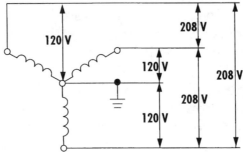

Figure 12-6 Wye system connections.

The center connection is considered to be the common connection and could be connected to the ground (electrical) and be used as a neutral wire. When each winding produces 120 V, there would be 208 V available between each pair of phase connections. In this type of system the alternator output voltage equals the winding voltage times the square root of three $\sqrt{3}$:

$$V_A = W_V \times \sqrt{3}$$

EFFECTIVE VALUES OF CURRENT AND VOLTAGE: The average values of ac electricity and dc electricity are quite different. We may have an ac electric circuit with an average value of 15 A flowing through it. However, this is no indication as to what the same current flowing in an identical dc circuit would do. Because more and more equipment manufacturers are including both ac and dc current in their equipment, it is well to know how the two values are related. With the use of effective values of voltages and current, this is possible.

The value of either voltage or current in an ac circuit that would produce the same amount of heat in a dc circuit containing only resistance when a dc voltage or current of the same value is flowing is the *effective value*. Thus, we know that an ac current with an effective value of 2 A will generate the same amount of heat in a 15-Ω resistor as a dc current of 2A. Because of the way it is derived, the effective value is also called the *root mean square (rms)*. The effective value equals the square root of the average value of the squares of all the instantaneous values of voltage and current during one-half of the cycle. The effective value if 0.707 E_{EFF} = 0.707I_{PK}. Therefore, if we have an ac voltage of 120 V, the effective voltage of E_{EFF} = 0.707 x 120 – 84.4 V. Thus, it we placed in a resistor in this circuit it would produce the same amount of heat as if we placed it in a dc circuit operating at 84.4 V.

Usually, ac voltage and currents are rated to provide the effective value. A voltage of 120 V is the effective (rms) value. The air conditioning and refrigeration technician will not be really concerned with these values, because the meters are designed to read the effective value of voltage and current. This material was presented for information and to increase your knowledge of the electrical theories in this text.

AVERAGE VALUES OF VOLTAGE AND CURRENT: The average values of ac voltage and current are the average values during one-half the cycle. See Figure 12-7.

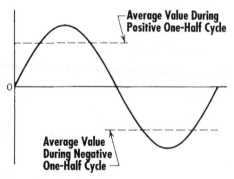

Figure 12-7 Average values of voltage and current.

In ac current generation the voltage and/or current increases, from zero point to the peak value; then decreases back to zero again. The average value must then, be at some point between zero and the peak value. For example, if we have a pure sine-wave circuit, the average value is 0.637 times the peak value. The voltage formula is E_{AV} = 0.637E_{PK} and for current I_{AV} = 0.637I_{PK}.

Example: What will be the average voltage in a circuit with a peak voltage of 120 V? $E_{AV} = 0.637E_{PK} = 0.637 \times 120 = 76.44$ V. *Note:* This value should be taken only on one-half the cycle because with a complete cycle this would be zero. This is because both halves of the cycle are identical, only one is positive and the other is negative.

FREQUENCY: The number of cycles (complete revolutions of the alternator) generated in one second is termed frequency of the current or voltage, and is expressed in hertz (Hz) per second. The frequency of ac voltage provides a means of classification. The power used in the United States is generally 60 or 400 Hz. The faster the alternations, the higher the frequency. See Figure 12-8.

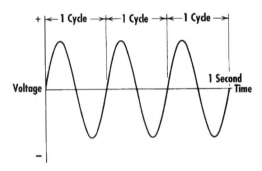

Figure 12-8 Frequency of electric power.

Equipment designed to be used in the United States has a label stating that 60-Hz electricity must be provided to the equipment. When the equipment is connected to a different type of electricity, damage will be done or the unit will not function as it was designed.

PHASE: The output of an ac alternator varies like a sine wave. A sine wave is represented by a full circle that has been cut at one point and one end held in place while the other is moved along an axis as if it were straightened out. See Figure 12-9.

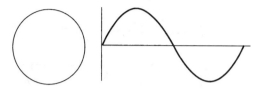

Figure 12-9 Sine wave from a circle.

This circle represents one complete revolution of the alternator. If two identical alternators were started at the same time and were turning at exactly the same speed, the sine wave will start and stop simultaneously. Their maximum values will be reached at exactly the same time as well as any other point in their sine wave. Thus, the two wave forms are said to be in phase. The term "phase" is thus used to indicate the time relationships between ac voltage and current. See Figure 12-10.

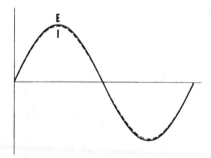

Figure 12-10 Current and voltage in phase.

Two currents and voltages can be in phase but their magnitude could be different. Their peaks could be reached at the same time but they can have different values. See Figure 12-11.

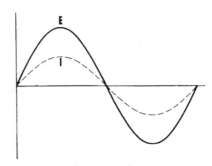

Figure 12-11 Current and voltage in phase with different values.

PHASE DIFFERENCE: A *phase difference* exists when the output of two alternators peak at different times. When one alternator is started after another, their maximum and minimum output values will occur at different times. Thus, the two outputs are out of phase, or there is a phase difference between the two alternator outputs. See Figure 12-12.

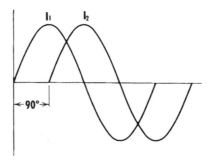

Figure 12-12 Phase difference.

The phase difference is stated in degrees of a circle for more accuracy.

When referring to phase difference, the terms lead and lag are used to describe the relationship between the two quantities that are being considered. To lead means to be in front of the other, while to lag means to be behind.

Unit 13: Electrical Distribution Systems

Almost all air conditioning and refrigeration equipment manufactured in the United States is designed to operate on ac current. They must be connected to the power distribution system through a service panel located on the premises of the user. The type used is determined by the type of service available. That is, 240 V single-phase, 240 V three-phase, 460 V three-phase, etc. The service and installation technicians must be able to determine which type is available.

GENERATING ELECTRICITY: When a current flows through a conductor there is a magnetic field produced around the conductor. When another wire passes through the magnetic field, a voltage is induced into that wire. The amount of voltage that is induced into the wire depends on two factors: (1) the strength of the magnetic field, and (2) the speed at which the wire passes through the magnetic field. Thus, the stronger the magnetic field, the higher the voltage, and the higher the speed of the wire, the higher the voltage.

An electric generator is a device that transforms mechanical energy into electrical energy by rotating a coil of wire in a magnetic field. See Figure 13-1.

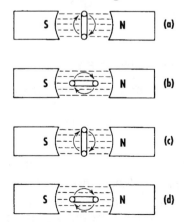

Figure 13-1 Action of an ac generator.

The coil of wire is called the *armature*. Notice the four positions of the coil during rotation. In Figure 13-1a, both wires of the coil are moving parallel to the magnetic field and since no line of flux are crossed, no voltage is induced into the moving coil. When the coil has moved to the position shown in Figure 13-1b, both sides of the coil are cutting across the magnetic field at right angles and the maximum voltage is induced into the coil. In Figure 13-1c, the coil position is opposite that in Figure 13-1a. The voltage is again reduced to zero. When the coil is rotated further, as in Figure 13-1d, the coil is again cutting the magnetic field at right angles and the maximum voltage is induced into the coil. Notice that the voltage generated in this position is opposite to that generated in the position shown in Figure 13-1b.

The voltage generated during this revolution of the coil can be drawn on a graph. See Figure 13-2. Remember, the strength of the induced voltage in a rotating coil is dependent on:

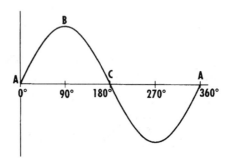

Figure 13-2 Voltage generated during one revolution of a generator.

1. The speed at which the wire passes through the magnetic field.

2. The number of magnetic lines of force cut by the moving coil.

When a single conductor moves across 100,000,000 lines of a magnetic field in 1 second, 1 volt of electricity will be produced. Thus, the armature may be wound with the correct number of turns to produce the desired voltage.

In actual practice, three rather than one or two windings are used in the ac alternator (or generator). In an alternator which is made with three windings the space between the windings is less and the load on the steam turbine or other source of energy is more consistent, thus reducing the vibration and other operating difficulties.

The windings are placed 120° (one-third of a complete circle) out of phase to each other. Thus, the output voltages are 120° out of phase. See Figure 13-3.

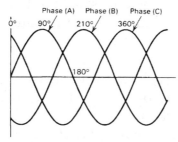

Figure 13-3 Three-phase voltage output graph.

TRANSMITTING ELECTRICITY: Electric power generating stations cannot always be located near the user, they are generally built near sources of natural energy such as large rivers, lakes, etc. The electricity must then be transmitted over great distances and made available to the users. If dc power is transmitted over long distances, large power losses would occur. These losses are reduced tremendously when ac power is transmitted. This is possible because the power can be transformed into a higher voltage and a lower amperage so that the heat produced by the resistance of the conductors is reduced as much as possible. Thus reducing the losses through heat being generated in the transmission process.

IDEAL CIRCUIT: If it were possible to build an ideal circuit, all the power produced by the alternator could be converted by the load into some useful purpose. Because it is impossible to build an ideal circuit, some of the energy generated is lost due to resistance inside the power source and the conductors that make-up the circuit. This is in the form of wasted energy and must be kept to a minimum. Most of the energy is wasted in the form of heat resulting when electrical current flows through the circuit.

LOSS OF POWER IN TRANSMISSION: Because the loss of energy due to the resistance of wiring to the flow of electric current certain steps must be taken in transmitting electricity over great distances. A certain portion of the electricity is converted to heat during the transmission at best. As stated earlier, this heat loss is directly proportional to the resistance of the circuit and to the square of the current. The formula for power loss is $P = I^2R$.

The losses in transmission are reduced by lowering either the current (I) in the transmission system, the resistance of the system, or both. It is better to reduce the current because it is squared rather than lowering the resistance, which has much less effect on power loss. Thus, in the transmission of ac electric power, the current carried by the system is relatively small.

When the end of the transmission lines is reached, the small current is changed into a larger current required by the users. Thus, the power loss (I^2R) is kept to a minimum. This increase is accomplished through the use of transformers placed in the transmission lines near the point of use.

It may seem confusing to discuss how electric power can be transmitted with a low current in the transmission line and have high current where the users need it. To help in understanding, remember the electric power equation for voltage and current: $P = EI$. By using this formula determine that an identical power (P) can be obtained from several different combinations of current (I) and voltage (E).

Transformers are the devices used to convert ac power from one combination of voltage and current to another combination. We will discuss transformers later in the text.

Unit 14: Distribution Center

The distribution center has many names. It is known as the circuit breaker box, the fuse box, the main load center, etc. Regardless of what name it is known by it is the part of the electrical circuit where the meter loop to the power distribution system is connected. The electric power is brought into the building at this point and connected to a main breaker. This main breaker must be heavy enough to permit the full load of the building to pass without tripping. See Figure 14-1.

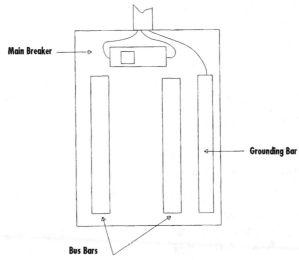

Figure 14-1 Electric power to the main breaker.

There will be either two or three power lugs and a ground (neutral) lug to which these wires are connected. The main breakers are mounted on a buss bar which allows the electricity to flow to the other circuit breakers. The ground wire is connected to the ground buss bar and is not protected by the circuit breaker. A single phase circuit breaker is used on single phase systems and a three-phase breaker is used on three-phase systems. The two types of breakers are not interchangeable.

Circuit breakers: circuit breakers are used to protect the circuit from an overload. When a breaker is rated for a higher current than the circuit, the wire could get overheated and cause a fire. Always use the proper size breaker or fuse for the circuit that it is to protect. Breakers are available is either single-pole, double-pole, and three-pole.

Single-pole breakers are designed to protect 120 V single-phase circuits. Double-pole breakers are designed to protect 240 V single-phase circuits, and three-pole breakers are designed to protect only three-phase circuits. Do not attempt to interchange their intended use.

BRANCH CIRCUITS: Branch circuits are those that carry the electricity to the point of use. They may be either 120 V single-phase, 240 V single-phase, or 240 V, 460 V or some other three-phase voltage.

When the circuit is being constructed there is usually at least three wires for each circuit. In a 120 V single-phase circuit there should be two power wires and a bare ground wire. A black wire is used for the "hot" wire and a white wire is used for the ground, or neutral wire. The bare wire is used for a ground wire. It is not a current carrying wire. The black wire will be connected to the circuit breaker or fuse outlet. The white wire will be connected to the grounding buss bar. The bare ground will also be connected to the grounding buss bar.

In a 240 V single-phase circuit there is usually two black wires and a ground wire. The ground wire may be either bare or white. However, when romex cable is used there will be one black wire and one white wire with the bare ground. When two black wires are used they are connected to the outlet of the 240 V breaker or fuse. The ground wire is connected to the grounding buss bar. When the romex cable is used both the black and the white wires are used as power wires. The bare wire is connected to the grounding buss bar.

In a three-phase system there are generally four wires to the circuit. Three of them are power wires and one is the ground wire. The power wires are connected to the outlet terminals of the breaker or the fuses. The ground wire is connected to the grounding buss bar.

At the unit, the power wires are connected to the power lugs on the unit. The ground wire is connected to a grounding lug located inside the unit cabinet. This wire is sometimes referred to as a mechanical ground.

FUSES: Fuses that are to be used to protect air conditioning and refrigeration equipment must be of the slow blow type. This type of fuse will carry an overload for a few seconds to give the unit time to reach its operating speed which causes it to draw a lower amperage. However, if the overload is not reduced sufficiently in the time frame for which it was designed it will blow and must be replaced before the equipment will operate again. Never jumper the fuse or circuit breaker or install one that is rated too high for the circuit. If there is continuous blowing of the fuse or tripping of the circuit breaker there is a problem that must be found and repaired. Simply increasing the size of the fuse or breaker will only cause more trouble later.

Unit 15: Single-Phase-60 HZ Voltage Systems

Most residences and small commercial buildings are supplied with single-phase power. Some of the larger residences may be supplied with three-phase, but they are rare. They require that single-phase air conditioning and refrigeration systems be installed. Most of the smaller units require single-phase power, regardless of where they are installed.

120 VOLT 60 Hz SINGLE-PHASE: The 120 V single-phase systems are used in residences to operate most of the appliances, lights, radios, televisions, and gas furnaces. These systems will normally have two wires with a bare ground. The black wire is used as the "hot" wire to feed electricity to the unit and is connected to the outlet connection of the circuit breaker or fuse. The white is the ground or neutral wire which is connected to the ground strip in the distribution center. The bare wire is the mechanical ground and is also connected to the ground strip in the distribution center. A single-pole breaker or a single fuse is used for these circuits. See Figure 15-1.

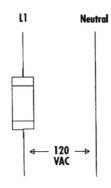

Figure 15-1 120-V single phase system.

The other ends of the wires are connected either to a plug or to the unit itself depending on the use of the power. The check for voltage can be done by touching one of the voltmeter leads to the black wire and the other meter lead to either the white wire or the bare wire. The meter will usually read around 117 to 120 V.

240 VOLT-60 Hz SINGLE-PHASE SYSTEMS: The 240 V-single-phase-60 Hz system is probably the most commonly used in the air conditioning and refrigeration industry. Therefore, it is necessary that the installation and service technicians be familiar with their operation. The installation technician must be able to determine what type of electric power is available so that the correct unit can be installed. The service technician must be able to determine what the of power is used so that the correct operating analysis can be made.

The 240 V single-phase 60 Hz system consists of three wires. There are two insulated wires and one bare wire included in the wiring to the unit. The size of the wire will, of course, depend on the current (ampacity) draw of the unit to which it is connected. If the wiring is in a romex there will be one black and one white insulated wires and one bare wire included. The insulated wires are connected to the outlet of a double-pole circuit breaker or the outlet connection of the fuse in the distribution center. The bare wire will be connected to the ground strip there also. The insulated wires will be connected to the power terminals at the unit and the bare wire will be attached to a grounding lug also at the unit. See Figure 15-2.

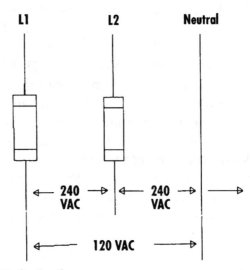

Figure 15-2 240-V single-phase system.

The control circuit to any type of system uses a single-phase transformer to change the applied voltage to 24 VAC to operate the controls. They will all be connected the same except that the 120 volt requires transformer with a 120 V primary and the 240 V system requires a transformer with a 240 V primary. A transformer with a 120 V primary can be used on a 240 V system by not connecting one of the power wires. See Figure 15-3.

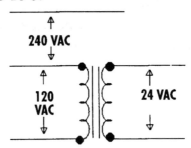

Figure 15-3 120-V transformer connected to a 240-V system.

However, it is not possible to connect a transformer with a 240 V primary to one having a 120 V power supply. Thus, the voltage to the primary of the transformer must be the same that the transformer is rated for. Otherwise the transformer will be ruined.

The 240 V system is usually equipped with a safety cut-off switch within reach of the unit. This is to allow the service technician to turn off the power when servicing the unit. There may or may not be fuses or circuit breakers in these cut-off switches. However, they are usually provided for safety purposes only.

Unit 16: Three-Phase Voltage Systems

Three-phase voltage systems receive three-phase power from the distribution system to the distribution center in the building. Three-phase voltage is commonly used in commercial and industrial buildings and some larger residences. Three-phase power has three power wires and one ground wire. This making it a four wire system. The three power wires are insulated and the ground wire may or may not be insulated. The installation and service technicians must be familiar with these types of systems so that they can be properly installed and serviced.

The power wires are connected to the circuit breaker or the fuse outlet connection in the distribution center of the building. The other ends are connected to the equipment power terminals. The ground, or neutral wire is connected to the grounding strip in the distribution center and to the ground provided at the equipment. See Figure 16-1.

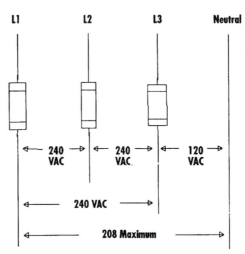

Wild leg may vary from 190 to 208 VAC to Neutral

Figure 16-1 240-V three-phase 4-wire system.

Three-phase power systems are more versatile than single-phase systems because of the different voltages that are available and some of the procedures that are used to properly use them. They also provide more torque for the equipment. There is no starting components required for the motors, thus making the troubleshooting much easier. Most of the large electric motors used in air conditioning and refrigeration are available in three-phase only. This is because of the better starting and running characteristics of three-phase equipment. However, three-phase distribution centers and equipment are more expensive than single-phase. Single-phase equipment can be operated from a three-phase system by not using one of the power wires. When this is method is used care must be taken not to use the same power wires for several single-phase units because this will cause an upset in the three-phase power and cause problems with it. Just make certain that the three-phase power remains in balance. That is, all the power wires carry the same amount of current. Also, the circuit breakers or fuses for the single-phase equipment must be sized for the load it serves. See Figure 16-2.

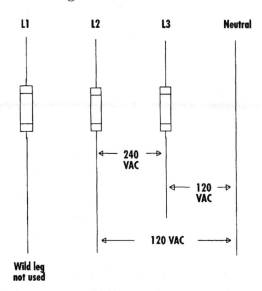

Figure 16-2 Reducing three-phase to single-phase.

240 V-THREE-PHASE-60 Hz "DELTA" SYSTEM: This type of system is usually supplied to buildings that require a large power supply to equipment used in the manufacture or treatment of materials in the building that require three-phase power for operation. The 240 V three-phase delta system uses four wires. It is called the delta system because of the connections of the transformer between the power distribution system and the meter loop into the building. It is shaped like the Greek letter delta (Δ). See figure 16-3.

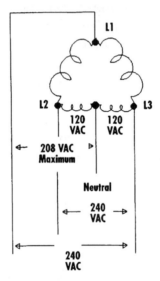

Figure 16-3 Connections for a 240-V-3-phase delta secondary.

There are three power wires and one ground wire. However, there are instances when it is supplied with only the three power wires. When the neutral wire is not used, the system cannot be reduced to 120 V single-phase because there is no ground wire. See Figure 16-4.

This system contains what is called a high leg. That is one of the power wires will have a higher voltage to ground than the other two. Usually the electrician will connect this wire to L_1 in the distribution center, or some type of orange marking. This higher voltage, or wild leg, is because the transformer winding is longer from L_1 to the neutral

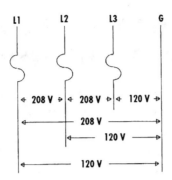

Figure 16-4 Schematic for a 208V-3-Phase-60Hz, four wire system.

connection than is L_2 or L3 to the neutral connection. This voltage will range from about 180 V to a maximum of 208 V. When measuring the voltage in this system from L_1 to L_2, L_2 to L_3, L_1 to L_3 will indicate that 240 V is present. When checking from L_2 or L_3 to ground or neutral indicates that 120 V is present. When any one of the three power wires is lost of a three-phase delta system, only single-phase power will be available. This is one condition that the technician must be aware of. A three-phase motor operating on single-phase power will burn-up from heat and loss of power. Some three-phase motors will continue to run if they are running when one of the power phases is lost.

The delta system is used mostly on installations that require a lot of 240 V three-phase circuits and only a few 120 V single-phase circuits. Only L_2 or L_3 to the neutral can be used to provide 120 V single-phase circuits. When L1 is connected to the neutral there will be from about 180 V to 208 V provided to the circuit. This type of voltage will damage any 120 V equipment connected to it. There are two methods of determining if a system is a delta: (1) by checking the voltage across any two of the three power wires. If 240 V is indicated it is a delta system, and (2) checking the voltage from each of the power wires to the neutral wire provides 180 V to 208 V.

Unit 17: 208 Volt–Three-Phase 60Hz "WYE" System:

The 208 V Wye system is generally used in installations where a large number of 120 V circuits are required. Usually in schools, hospitals, office buildings, and other such buildings. This type of system provides the three-phase power needed to operate heavy equipment and can also be divided into 120 V circuits for operating smaller equipment such as lights, Small refrigeration systems, and other types of specialty equipment that operates on 120 V single-phase power. When using three-phase equipment in this type of installation, be sure to use 208 three-phase or 208-240 V three-phase equipment. The lower voltage is enough to cause the straight 240 V equipment to operate on low voltage which will cause many operating problems.

The 208 V three-phase system is supplied to the distribution center by three power wires and one neutral ground wire. See Figure 17-1.

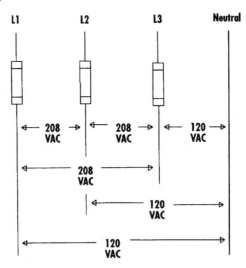

Figure 17-1 208-V, three-phase, 4-wire system.

The 208 volt system has no wild leg. L1, L2, and L3 are power wires and G represents the ground or neutral connection.

When checking the voltage between any of the three power wires 208 volts will be indicated. When the voltage between any one of the power wires is tested to ground 120 V will be indicated. Thus, there are three 120 V power legs available. This will allow more 120 V circuits than the 240 V delta system without becoming unbalanced.

The connections to the secondary of a 208 V three-phase Y system is different from the 240 V three-phase delta system. See Figure 17-2.

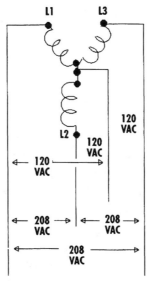

Figure 17-2 *Connections for a 208-V three-phase-60 Hz Wye secondary transformer.*

The transformer windings in the wye transformer are shaped like the letter Y from which the system gets its name. Notice that all three of the transformer windings are the same length. This type of transformer connection is used on some of the higher voltage systems.

There are two methods that may be used to determine if it is a wye system: (1) the voltage between any two of the three power legs will be 208 volts, and (2) the voltage between any one of the power legs and the ground will be 120 volts.

Unit 18: Higher Voltage Systems

There are many advantages in using voltages higher than 208, and 240 V. These types of systems are generally used in industrial and commercial applications. The technician must be able to determine what type of system is being worked on for safety and to prevent equipment damage.

Higher-voltage systems include those that provide 240/408 V-single phase, 240-416 V-three-phase, and 277/480 V-three-phase power to the structure. It is relatively simple to determine which system is being worked with by checking the voltage with an voltmeter. In the 277/416 volt-three-phase system, a 277 volt reading can be found when reading between any of the hot legs and the ground. These types of systems are available in either delta or wye connections, depending on the locality. The 277/480 volt wye system is the most popular of the two systems.

SYSTEM ADVANTAGES: The first advantage to using the higher voltage systems is because the wiring used is rated at 600 volts. Second, the system components, such as the relays, switches, and electric panels are the same as those used on 208-volt and 480-volt type systems. The motors used on higher-voltage systems are generally wound to operate on more than one voltage. The wiring and the service entrance equipment will usually have a lower first cost to the customer.

DISADVANTAGES: When 120 and 240 volt single-phase systems are required, additional transformers will need to be purchased to meet these requirements. This will add to the first cost to the customer. The motors are more costly because they must be wound to accommodate more than one voltage, which is more costly. And, single-phase equipment is more popular than three-phase.

Unit 19: The Meter Loop

The meter loop is the part of an electrical system that extends from the electric pole, through the meter and into the service panel, or circuit breaker box. See Figure 19-1.

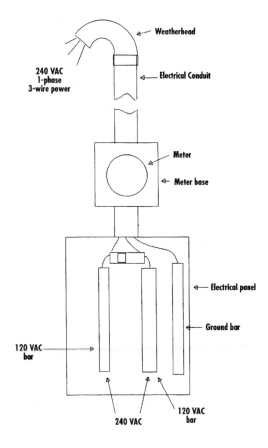

Figure 19-1 The single-phase meter loop.

The power brought into the structure may be either single-phase of three-phase depending upon the requirements of the user. The power from the pole may be either three wire of four wire. When the electricity passes through the meter, the power used is measured and registered on the meter. This is how the power company determines the charge to the customer. As the power leaves the meter, it is connected to the proper terminals inside the breaker box, or distribution center, so that the desired type and amount of circuits can be wired by the electrician. See Figure 19-2.

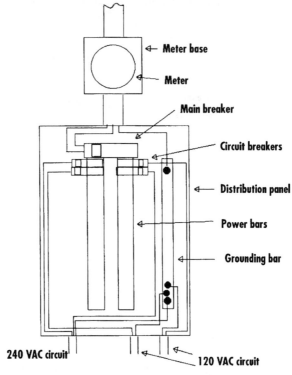

Figure 19-2 Connections to the Distribution center.

Some power companies refer to the meter loop as being that part from the weatherhead to the distribution center as being the meter loop. However, this is a matter of terminology and power company policy.

Unit 20: Circuit Protection Devices

Circuit protection devices are those that are designed to protect a circuit from overload. Should an overload occur in the circuit, the protective device will interrupt the power to that circuit. Circuit protective devices generally are fuses and circuit breakers.

FUSES: Fuses are available in either the cartridge type or the plug type. The cartridge type is usually more popular on larger equipment. The plug type fuse is generally used on furnaces. The circuit fuse is located either in the distribution center of in a fused disconnect. See Figure 20-1.

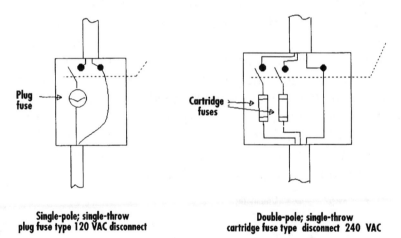

| Single-pole; single-throw plug fuse type 120 VAC disconnect | Double-pole; single-throw cartridge fuse type disconnect 240 VAC |

Figure 20-1 Fused disconnects (plug and cartridge).

They are sized for the maximum ampacity (amperage) of the circuit. When this value is exceeded the fuse will blow, disconnecting the circuit from the distribution center.

Cartridge type fuses are available in the replaceable link type or the one time type. The size of the fuse is determined during the manufacturing process by placing an element (link) inside the fuse that will only carry a given amount of current. See Figure 20-2.

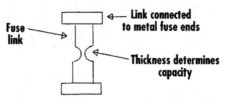

Figure 20-2 Inside a fuse.

Plug type fuses are round and have a window that will usually indicate if the fuse is blown. The window will appear to be burned and broken. When a fuse is suspected of being blown always check it.

Fuses are very easily checked. Simply turn off the power to the fuse box and remove the fuse. Check the circuit through the fuse with an ohmmeter. If the fuse is good it will usually show zero ohms (no resistance) on the ohmmeter. When it is bad is will show infinity. However, sometimes a fuse will not blow completely. It will have sufficient connection to indicate zero ohms on the meter but it will not allow enough current to flow through it to operate the equipment. If the fuse appears good and shows continuity, but the equipment still will not work, place the fuse back into its holder and apply voltage to it. Then with the voltmeter, check the voltage across it to determine if there is a voltage drop at this point. See Figure 20-3.

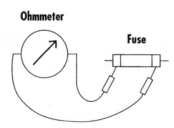

Figure 20-3 Checking a fuse.

Be sure to change the meter from ohms to volts to prevent damage to the instrument.

When a fuse blows there is a reason. The complete circuit must be checked to locate the overload and the problem corrected. Otherwise the fuse will continue to blow, which will cause a very dissatisfied customer. Do not increase the size of the fuse to prevent it from blowing. Find the problem and repair it.

CIRCUIT BREAKERS: Circuit breakers have the same responsibility as the fuse. Its purpose is to protect the circuit from overload and damage. Breakers are placed in the distribution center. They are sized for the maximum ampacity of the circuit. This is usually indicated by the equipment manufacturer.

Circuit breakers are manufactured with a heat sensing element inside their case. See Figure 20-4.

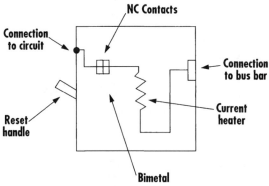

Figure 20-4 Inside a circuit-breaker.

When the current draw exceeds their rating, the heat will cause the circuit contacts to open and interrupt the power to the circuit. When they trip the handle usually moves to a position indicating that they are open. To reset a breaker simply move the handle to the complete off position and then move it back to the closed position. Check the current draw through the breaker to determine if the circuit is overloaded. If there is and overload, the problem causing the overload must be found and corrected.

If the current draw is within the rating of the breaker, allow the unit to operate for a time while checking the amperage draw through it. If the amperage draw does not increase but the breaker trips again, the breaker is weak and must be replaced. Do not install a breaker with a higher rating than the one being replaced because damage to the circuit may occur.

Unit 21: Inductance And Reactance

Inductance may be defined as that property in an electric circuit that resists a change in the flow of current. This opposition to change occurs because of the energy stored in the magnetic field of an energized coil. This field strength is measured in henries and is indicated by the letter L. Inductance is caused by self-induction of the coil when current is flowing through it. The expanding and collapsing of the magnetic field is what causes the field to continually move across the conductor, or wire, from which the coil is made. See Figure 21-1.

Magnetic Field

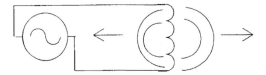

Fig. 21-1. The magnetic field expanding.

The induced voltage is opposite to the applied voltage. Because of this, inductance opposes every change of current in the coil. See Figure 21-2.

Magnetic Field

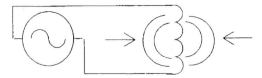

Fig. 21-2. The collapsing magnetic field.

This induced voltage is commonly called counter electromotive force of CEMF or back EMF. One henry of inductance is present when one volt of induced EMF is produced when the current flow changes at the rate of one ampere per second. Mathematically this is:

E	=	L x (ΔIΔt)
where: E	=	induced voltage
L	=	inductance, in henries
ΔI	=	change of current flow, in amperes
ΔL	=	change in time, in seconds

The symbol Δ means "a change in." The amount of induced voltage depends on the strength of the magnetic field inside the coil. When a strong magnetic field is present a higher induced voltage is produced. A strong magnetic field has more lines of force than a weaker one. It is the number of lines of force that cut across the conductor that determines the strength of the CEMF. The strength of the magnetic field can be increased by inserting a soft iron core inside the coil. See Figure 21-3.

Metallic Core

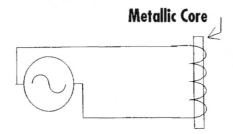

Fig. 21-3. Magnetic coil with a metallic core.

The stronger field is possible because the soft iron core has a higher permeability than the air which is normally inside the coil. The magnetic field will pass through the core much easier than it will through the air, thereby concentrating it into a smaller space. Permeability is a measure of a material to pass magnetic lines of force.

Thus, we can see that inductance is the CEMF induced in a conductor each time a magnetic line of force cuts through a conductor. Due to the time lag between the time the magnetic field cuts the conductor and the CEMF is generated, the CEMF is 180° out of phase with the applied voltage, or exactly opposite to it. This induced voltage also has a current-limiting characteristic which acts much like a resistance, however, it must not be be considered as resistance, but as reactance, in the circuit.

Inductors, or wire wound, electric coils are used in many applications in air conditioning and refrigeration applications. The action of the CEMF produced in electric motors keeps the motors from continually speeding up and flying apart from the centrifugal force on the rotating parts.

Self-Induction: Self-induction occurs in every circuit that has current flowing through it. A change of current flow through an inductor causes the magnetic field to expand and contract. The lines of force of the magnetic field cut across the conductor and a CEMF is produced. The CEMF is exactly opposite to the applied voltage and opposes any change in current. This is called self-induction. The strength of the self-inducted energy depends on the number of turns in the coil, the relationship between the length of the coil and its diameter, and the permeability of the core material.

Mutual Induction: Mutual induction may be described as the ability of one electric circuit to induce a voltage into another, even when the two inductors are electrically separated. See Figure 21-4.

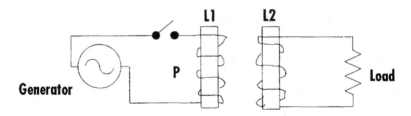

Figure 21-4: Mutual Induction.

Usually mutual induction is associated with two conductors, however, mutual induction can actually occur between two straight conductors that are close together. However, this induction will not be very strong because there is not enough magnetic lines of flux cutting the conductor. Mutual inductance actually is the transfer of electricity from one circuit into another separate circuit through the magnetic field created when current flows through a circuit.

When two coils are placed close enough so that their magnetic lines of force combine with each other, mutual inductance occurs. When the coils are placed close together the induced voltage will be stronger. When it is possible to use the same core for both coils will cause a stronger induced voltage also. The degree to which the magnetic lines of force from one coil cuts across the windings of the second coil is called coupling. When all of the lines of force cut across all of the turns of wire in the second coil, unit coupling is accomplished.

Inductive Reactance: When a circuit contains only inductance, the amount of current that will flow through the circuit is determined by the strength of the CEMF. The inductive reactance of a circuit is measured in ohms and is indicated by X_L. The CEMF, which is caused by an inductor, is controlled by the inductance (L) of the inductor and the frequency (f) of the applied current, the inductive reactance also depends on the same factors. Inductive reactance can be calculated by using the equation:

$$X_L = 2\pi fL$$

where: X_L = inductive reactance, in ohms
2π = frequency of current, in hertz
L = inductance, in henries

We can determine from the formula that the higher the frequency or the greater the inductance, the greater will be the inductive reactance. Also, the lower to frequency or inductance, the smaller the inductive reactance.

Unit 22: Capacitance And Capacitors

Capacitance and inductance are two properties in an ac electrical circuit that affect its operation. Both of these properties resist a change in a particular phase of current flow. Capacitance resists any change in voltage and inductance opposes any change in current.

Capacitance may be defined as that property in an electrical circuit which opposes any change in voltage. Capacitors are used to store electrical energy through the use of an electrostatic field and to release this energy back into the circuit sometime later.

The capacitance of a capacitor is determined by the number of electrons that may be stored on its plates for each volt of applied voltage. Capacitance is measured in farads, which represent a charge of one coulomb, which raises the electrical potential one volt. To calculate capacitance use the following formula:

C	=	Q ÷ E
Where C	=	farads
Q	=	coulombs
E	=	volts

The capacitors used in air conditioning and refrigeration systems are measured in microfarads (µF; one millionth of a farad) because one farad is an enormous amount of electrical charge. The Greek letter µ is used to represent micro. See Table 22-1.

Conversion of Units
1 farad (f) = 1,000,000 microfarads (µf) = 1,000,000,000,000 picofarads (pf)
1 microfarad (µf) = $\frac{1}{1,000,000}$ farad (f) = 1,000,000 picofarads (pf)
1 picofarad (pf) = $\frac{1}{1,000,000,000,000}$ farad (f)
1 picofarad (pf) = $\frac{1}{1,000,000}$ microfarad (µf)
1 picofarad (pf) = 1 micromicrofarad (µµf)

Table 22-1 Conversion of units.

The capacitance of a capacitor is determined by the following factors:

1. The material used as the dielectric (insulator)

2. The surface area of the plates

3. The distance between the plates

The capacity of a capacitor increases when the surface area of the plates is increased or when the dielectric is increased. Also, the capacitance decreases when the distance between the plates is increased.

CAPACITORS: The capacitor is a simple component. It may be described as two plates of a conductive material separated by an insulator, or dielectric. See Figure 22-1.

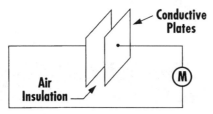

Figure 22-1 Simple capacitor.

The capacitor is shown as two metal plates separated by air insulation. If this circuit were connected to a dc power source, the circuit would appear to be open because the plates are not physically touching. See Figure 22-2.

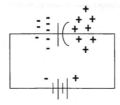

Figure 22-2 Simple capacitor in a dc circuit.

If we placed an ammeter in the circuit, it would show only a momentary flow of current when the switch is closed. Note that when the switch is closed, electrons will flow from the negative terminal of the power source to one plate of the capacitor. These displaced electrons will repel the electrons from the other plate of the capacitor, which will be attracted to the positive terminal of the power source. The capacitor and the power source now have the same electrical potential. However, these two charges are opposite in polarity. If the capacitor were removed from the circuit, it would maintain this charge because the energy is stored within its electric field.

Note that in this circuit no electrons flowed through the capacitor. A capacitor will block the flow of direct-current electricity. One of the plates will become negatively charged, the other one will be positively charged, creating a strong electric field between the two plates. The dielectric materials used in manufacturing capacitors vary in their ability to support an electric field. This ability is called the dielectric constant of the material. The dielectric constants of some familiar materials are listed in Table 22-2.

MATERIAL	DIELECTRIC CONSTANT
Air	1
Wax paper	3.5
Mica	6
Glass (window)	8
Pure water	81

Table 22-2 Dielectric comparison of some materials.

The constants of various materials used as dielectrics are compared to dry air, which is assigned the value of 1.

However, in an alternating-current circuit the applied voltage and current periodically change direction. Therefore, a capacitor placed in an ac circuit is charged first by the voltage applied in one direction, then in the opposite direction. When the applied voltage reaches a maximum value and starts to decrease, the flow of current also decreases; however, the capacitor is still being charged by voltage in the same direction. As the applied voltage continues to drop, the voltage that is developed across the capacitor plates becomes greater. The capacitor, now acting as the power source, starts discharging back into the circuit. See Figure 22-3.

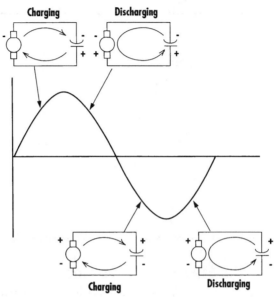

Figure 22-3 Capacitor d/charging and discharging sequence.

When the applied voltage reaches zero and reverses direction, the capacitor is considered to be fully charged. As the voltage passes zero the capacitor starts charging again but in the opposite direction from that in which it was previously charged. This charging action continues until the applied voltage starts dropping again in value and the charging and discharging process is repeated.

Remember that the charging path for a capacitor in an ac circuit is from the negative terminal of the power source to one of the capacitor's plates and from the other capacitor plate to the positive terminal or the power source. The discharging path for a capacitor is from the negative capacitor plate through the power source and back to the positive capacitor plate. It should be remembered, though, that in an ac circuit the negative and positive plates refer to a specific instant of time because of the alternations of the current.

RELATIONSHIPS OF VOLTAGE: The electrical charge on a capacitor is equal to but exactly opposite to the applied voltage. Thus, they are out of phase with each other. See Figure 22-4.

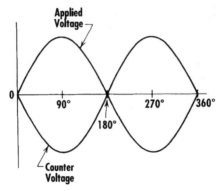

Figure 22-4 Applied voltage and counter voltage (CEMF).

Because the capacitor will develop its own voltage, it acts as a power source during the cycle and attempts to force current to flow into the power source. The voltage developed by the capacitor is therefore called a *counter voltage* because it is 180° out of phase with the applied voltage.

RELATIONSHIP OF VOLTAGE AND CURRENT: If a source of ac voltage is connected across a capacitor, the maximum amount of current would flow in the circuit the instant the applied voltage begins its sinusoidal rise from zero. This is because the plates are neutral in charge and offer no opposition to the flow of current. However, as the power source voltage increases, the capacitor begins to take on a charge because of the current flow. The capacitor voltage then offers and increasing opposition to the lower voltage and the current flow decreases. As the power source voltage reaches its maximum value, the voltage across the capacitor plates also reach a maximum value. This amount of charge is strong enough to cancel the applied voltage completely and the current flow in the circuit is stopped. As the applied voltage decreases toward zero, the charge on the capacitor plates become stronger than the power source voltage and the capacitor keeps discharging. The current through a capacitor leads the applied voltage across the capacitor by 90°. See Figure 22-5.

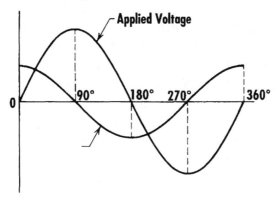

Figure 22-5 Applied voltage and current relationship.

Because the counter voltage is 180° out of phase with the applied voltage, the current lags the counter voltage by 90°. See Figure 22-6.

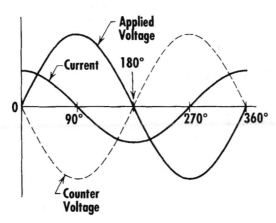

Figure 22-6 Applied voltage, counter voltage, and current relationship.

CAPACITANCE AND CAPACITOR CURRENT: Recall that capacitance is the amount of charge that can be stored in a capacitor for each volt applied to the capacitor. The formula for calculating capacitance is Q = CE. If two voltages of equal value are applied to capacitors with unequal values in capacitance, the electrical charge stored in the capacitors will be different. Capacitors with larger capacitance ratings will store more energy than smaller-rated capacitors. If the charging time was identical for both capacitors, as would be the case if the applied voltages and frequencies were the same, the current through the larger-rated capacitor would be greater. Thus, the current flow through a capacitor is directly proportional to the capacitance.

CAPACITORS IN SERIES: Connecting capacitors in electrical series has the same effect as increasing the distance between the plates of a single capacitor. The total capacitance, then, becomes less than the capacitance of the smallest capacitor. To calculate the value of capacitors in series, use the following formulas: $C_t = C_1 \times C_2 \div C_1 + C_2$ for two capacitors of equal value, or $1 \div C_t + 1 \div C_1 + 1 \div C_2 + 1 \div C_3$ ... for capacitors of unequal value. For example, see Figure 22-7, in which

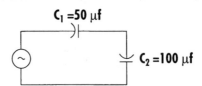

Figure 22-7 Calculating series capacitance of equal-value capacitors.

$C_t = C_1 \times C_2 \div C_1 + C_2 = 50 \times 100 \div 50 + 100 = 5000 \div 150 = 33.33\mu F$

Or See Figure 22-8, in which

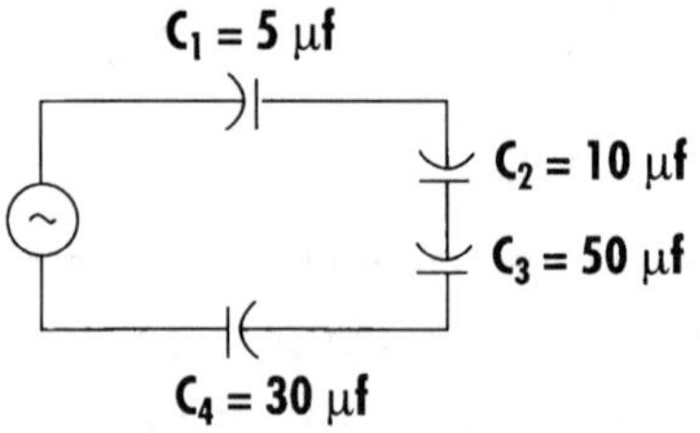

$C_1 = 5 \mu f$

$C_2 = 10 \mu f$

$C_3 = 50 \mu f$

$C_4 = 30 \mu f$

Figure 22-8 Calculating series capacitors of unequal values.

$1 \div C_t = 1 \div C_1 + C_2 + C_3 + 1 \div C_4 = 1 \div 5 + 1 \div 10 + 1 \div 50 + 1 \div 30 = 30 + 15 + 3 + 5 \div 150 = 53 \div 150 = 2.83\mu F$

CAPACITORS IN PARALLEL:

When capacitors are connected in parallel, the result is the same as enlarging the plate area of one capacitor. The total plate area of each capacitor is exposed to the total source voltage. Thus, when capacitors are connected in parallel, the total capacitance is increased. Therefore, the total capacitance of capacitors in parallel is calculated by the formula $C_t = C_1 + C_2 + C_3 + ...$

As an example of calculating total capacitance in a parallel circuit, see Figure 22-9, in which $C_t = 50 + 100 + 500 = 650\mu F$

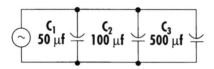

Figure 22-9 Calculating capacitors in parallel.

CAPACITANCE IN AC CIRCUITS: Capacitors, when subjected to alternating-current power, have a reactive quality. This quality is called *capacitive reactance*. Capacitive reactance is an opposition to a change in current. The recognized symbol for capacitive reactance is X_c, measured in ohms. The amount of capacitive reactance (X_c) in a given capacitor depends on its capacitance and the phase angle of the applied voltage. Capacitive reactance is the reciprocal relationship

X_c	=	$1 \div 2\pi fC$
where X_c	=	capacitive reactance
2π	=	a constant
f	=	voltage source frequency, in hertz (cycles per second)
C	=	capacitance of the capacitor

As an example, see Figure 22-10.

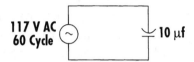

Figure 22-10 Calculating capacitive reactance.

Example: What is the capacitive reactance of a 10-μF capacitor operating in a circuit having a frequency of 60 Hz?

$X_C = 1 ÷ 2\pi fC = 1 ÷ (2 \times 3,14)(60)(10) = 1 ÷ 3768 = 0.0002656\Omega$

Once the capacitive reactance has been determined, it may be handled as any other opposition to current flow, with one exception . The applied voltage and current are 90° out of phase, with the current leading the voltage. See Figure 22-11.

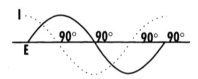

Figure 22-11 Phase relationship of capacitive current and voltage.

During the first instant when the voltage is applied to a capacitor, there is no opposition to current flow. The current then jumps almost instantly from zero to the peak value. Almost immediately, electrons begin to accumulate on one plate while the other plate becomes deficient in electrons. This charge has a polarity opposite to that of the power source. Since the charge increases on the plates, the opposition to the flow of current also increases. As the opposition increases, the current decreases. The current flow reaches aero at the same time the voltage reaches the peak value.

REACTIVE POWER: As a capacitor is discharged, the energy stored in the dielectric is returned to the circuit. This can be compared to inductance, which also returns the stored energy in the magnetic field to the circuit. In both cases the electrical energy is used only temporarily by the reactive circuit. The power in a capacitive circuit is called *wattless power*. The power-wave form shows that equal amounts of positive power and negative power are used by the circuit, thus resulting in zero power consumption. Therefore, the true power is zero. The apparent power is equal to the product of effective voltage and the effective current. An applied alternating-current (ac) voltage of 100 V to the capacitive circuit causes a current flow of 5A. The apparent power will be 100 V x 5 A = 500 VA. See Figure 22-12.

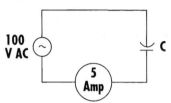

Figure 22-12 Theoretical capacitive circuit.

The relationship that exists between the true power and the apparent power in an ac circuit is called *the power factor*.

Power Factor (PF) = 0 true power ÷ 500 VA (apparent power)

The power factor may also be found by trigonometric equations. The power factor is equal to the cosine of the phase angle between the current and the voltage. The current leads the voltage by an angle of 90° and the phase angle is therefore 90° in a pure capacitive circuit.

The true power may be found by the following formula:

TP (true power) = I x E x cos = 10 x 100 x cos 90° (cos 90° = 0) = 0.

Unit 23: Magnetism

The effects of magnetism has been of great interest to scientists for centuries. The Greeks discovered the principle more than 2000 years ago. It was noted that a certain kind of stone was attracted to iron. Later the small stones were determined to be iron ore. These small stones were called magnetite, because they were usually found near Magnesia, in Asia Minor. The stones then became known as "leading stones" or loadstones because they were used by mariners for navigation purposes. They were the first natural magnets. When they were suspended by a string they would search out the north and south poles of the earth. See Figure 23-1.

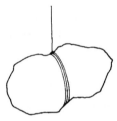

Figure 23-1 Loadstone.

Scientists then discovered that magnetism had some properties that resembled the effects of electricity. Sometime later the discovery was made that proved that the magnetic properties electric properties actually complimented each other. Today electricity and magnetism are so entwined that they can hardly be separated. Magnetism is used in the production of electricity (EMF); when electrons flow through a conductor, a magnetic field is present around the conductor. Thus, when you learn something about one it only enhances your knowledge of the other.

MAGNETISM AND ELECTRONS: As stated before the forces of electricity and magnetism are very closely related; however, each one is completely different from the other. When there is no motion present, the magnetic forces and electrostatic forces have no effect on each other. However, when either field is set in motion something happens that causes an interaction between the two forces. Because the electron is the smallest particle of matter, scientists have developed a theory to help explain the relationship of magnetism and electricity which is known as the *electron theory of magnetism.*

We learned earlier that an electron has a negative electrical charge, and that the lines of force caused by this charge come straight into the electron from all directions. See Figure 23-2.

Figure 23-2 Electrostatic field on an electron.

Scientists claim that because of its orbital spin, an electron also produces a magnetic field that forms concentric circles around the electron. See Figure 23-3.

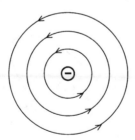

Figure 23-3 Magnetic field of an electron.

This theory indicates that the electrostatic lines of force and the magnetic lines of force are at right angles to each other. These combined lines of force are known as the electromagnetic field. See Figure 23-4.

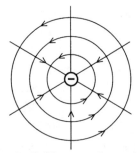

Figure 23-4 Electromagnetic field around an electron.

THE MOLECULE AND MAGNETISM: The only known natural magnetic metals are iron, nickel, and cobalt. This is because electrons tend to pair off in orbits having opposite spins. Their magnetic fields of force are also opposite and tend to cancel each other out. This does not mean that those elements that have an odd number of electrons are magnetic. However, if we could isolate the odd electrons, we would have a magnet. When atoms combine to form molecules--in all metals except iron, nickel, and cobalt--they tend to arrange themselves to have a total of eight valence electrons. In the arrangement process the orbital spins of the electrons cancel the magnetic field in most metals. See Figure 23-5.

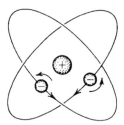

Figure 23-5 Magnetic atom.

Ions are formed when iron, nickel, and cobalt atoms combine which share their valence electrons so that many of the electron spins are not canceled, but are added together. This alignment of molecules produces regions in the metal that are known as magnetic domains. These molecules are then known as *magnetic molecules*, and act just like any other magnet. See Figure 23-6.

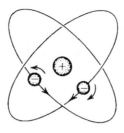

Figure 23-6 Magnetic molecule.

LAWS OF MAGNETISM: A magnet can attract iron. When two bar magnets are hung on strings so that they can rotate, we can see how the poles react to one another. See Figure 23-7.

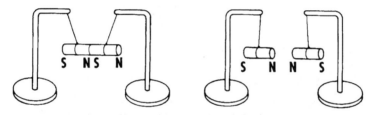

Figure 23-7 Demonstrating the laws of magnetism.

When the N (north) pole of one of the magnets comes close to the S (south) pole of the other magnet, they are attracted by a magnetic force that will cause the two magnets to move closer together. However, if the magnets are turned so that either the two S (south) poles or the two N (north) poles are close together, a repulsive magnetic force between them will cause the magnets to move away from each other. Thus, the *laws of magnetism* are: unlike poles attract each other and like poles repel each other.

There is a third law of magnetism that states that that attractive force increases as the distance between the two magnetic poles of the opposite polarity is decreased. This magnetic force, either attractive or repelling, varies as the square of the distance between the magnetic poles. For example, if the distance between two like magnetic poles is increased to two times the distance, the repulsive force will be reduced to one-forth of its original value.

MAGNETIC FIELDS: There is an invisible field around a magnet that resembles the electrostatic field. A *magnetic field* is the area around a magnet where its influence can be felt. This force field has a definite direction and its strength can be measured.

If we placed two magnets under a piece of paper that has iron filings on the top, the invisible fields can be seen. See Figure 23-8.

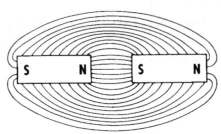

Figure 23-8 Attracting magnetic fields.

Notice that when the N and S pole are placed together the flux field is very strong. However, when the magnets are placed with like poles together, the repelling force is indicated by the absence of flux lines between the two magnets. See Figure 23-9.

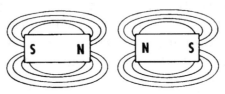

Figure 23-9 Repelling magnetic fields.

It should also be noted that the flux lines never cross, regardless of how dense they are. AS the distance between the two magnets is increased the flux lines become farther apart.

MAGNETIZING MATERIALS: Basically, there are two ways to make a magnet: **(1)** stroking with another magnet, and **(2)** using an electric current. When making a magnet the material being magnetized must be ferromagnetic.

Ferrous means iron, and each of the materials act like iron with respect to magnetism.

When using one magnet to make another out of a ferromagnetic material by stroking, the metal must be stroked in the same direction using the same pole of the magnet. This action will cause the particles of the material to align themselves in the direction of the strokes. To make a magnet, the particles of the material must already have magnetic properties. However, the overall magnetic properties of a piece of metal are neutral because the poles of the individual particles are arranged in a random pattern. See figure 23-10.

Figure 23-10 *Random arrangement of molecular poles.*

The repeated stroking action with another magnet will cause the molecules to align themselves. See Figure 23-11.

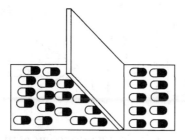

Figure 23-11 *Creating a magnet by stroking.*

If an unmagnetized piece of iron is placed inside a coil of electric wire and the wire is connected to a battery, the electric current flowing through the wire causes a magnetic field around the metal causing the molecules to align themselves and the iron is magnetized. See Figure 23-12.

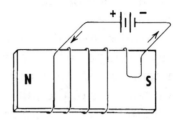

Figure 23-12 Creating a magnet with an electric current.

ELECTRIC CURRENT AND MAGNETISM: The relationship between electricity and magnetism has been attracting the interest scientists for centuries. A Danish physicist, Hans Christian Oersted, discovered that a conductor carrying an electric current is surrounded by a magnetic field. If small compasses were placed on a piece of cardboard close to the current carrying conductor which has a magnetic field surrounding it, the compass needles would point in the direction of the magnetic lines of force. See Figure 23-13.

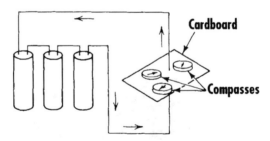

Figure 23-13 Compasses indicating direction of magnetic flux.

Should the direction of the current be reversed, the compasses will also be reversed by 180°, indicating that the direction of the magnetic field around a current-carrying conductor depends on the direction of current flow.

Unit 24: Electromagnetism

Static electricity causes no magnetic field. This is because with static charges, electrons with opposite spins pair off and cancel their magnetic effects.

A magnetic field is caused by current flowing through a conductor. The electrons flowing through a conductor cannot pair off with opposite spins. However, because they are flowing in the same direction their magnetic fields tend to add, increasing their strength. The magnetic field around a conductor carrying an electric current is in concentric circles around that conductor. See Figure 24-1.

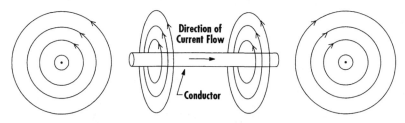

Figure 24-1 Magnetic lines of force around a conductor.

MAGNETIC FIELD STRENGTH: The greater the amount of current flowing through the conductor, the greater the strength of the magnetic field around that conductor. The magnetic lines of flux that are closest to the conductor are closer to each other. See Figure 24-2.

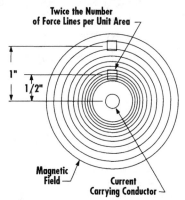

Figure 24-2 Magnetic flux lines around a conductor.

As the distance from the conductor is increased, so does the distance between the flux lines. From this it can be determined the flux lines that are closest to the conductor have the most strength.

The number of flux lines per unit area is in inverse proportion to the distance from the conductor. For example, at a distance of 1 inch from the conductor the field will have only one-half the density of that which is had 1/2 inch from the conductor.

ELECTROMAGNETIC FIELD INTERACTION: When two conductors in parallel are carrying current in opposite directions their magnetic fields will also be in opposite directions and will oppose each other. Remember, the flux lines cannot cross each other; if the two conductors were brought closer together, the force fields would tend to push the conductors apart. See Figure 24-3.

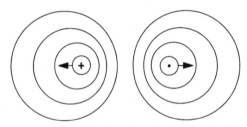

Figure 24-3 Interaction of opposite electromagnetic fields.

On the other hand, when two conductors are carrying current in the same direction are brought close together, their magnetic fields would aid each other, because the flux lines are in the same direction. Thus, the conductors would tend to be pulled closer together. See Figure 24-4.

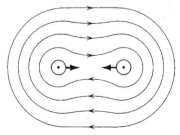

Figure 24-4 Interaction of aiding electromagnetic fields.

124

The two flux fields are strengthened by each other. When more conductors are used a stronger electromagnetic field will be caused, because all the fields aid each other. See Figure 24-5.

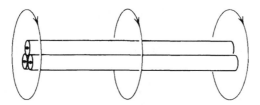

Figure 24-5 Combined wires produce stronger fields.

ELECTROMAGNETIC FIELD IN A LOOPED CONDUCTOR: When a current-carrying conductor is coiled into a loop, all the magnetic lines of flux around the conductor will flow out of the loop on the same side and enter on the same side. See figure 24-6.

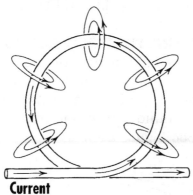

Current

Figure 24-6 Magnetic flux lines around a loop.

Thus, in the center of the loop the flux lines are more dense, causing a stronger magnetic field. See Figure 24-7.

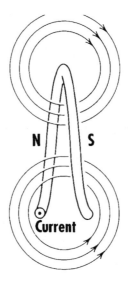

Figure 24-7 Magnetic flux lines in the center of the loop.

This force causes magnetic poles in the coil, with the N pole being on the leaving side and the S pole being on the entering side.

ELECTROMAGNETISM IN A COILED CONDUCTOR: In the previous example, if the conductor was extended to form a coil, the fields will support each other and make the flux lines through the coil stronger. When several loops of wire are wound around a core, the magnetic field includes the entire coil. See Figure 24-8.

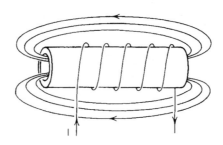

Figure 24-8 Magnetic field around a coil.

THE SOLENOID: When a conductor carrying current is wound into a solenoid, or coil, the individual magnetic lines of force around the conductor tend to join together.

The strength of the magnetic field created by a solenoid depends on two things: **(1)** the number of turns of wire in the coil, and **(2)** the current in amperes flowing through the wire. The product of the amperes and the number of turns is referred to as the ampere-turns of a solenoid; these indicate the strength of the magnetic field. For example, if a solenoid of 100 ampere-turns will produce a certain field strength, any combination of turns and amperes with a product of 100 will be satisfactory, for example:

10 turns x 10 amperes	=	100 ampere-turns
50 turns x 2 amperes	=	100 ampere-turns
100 turns x 1 ampere	=	100 ampere-turns

MAGNETIC CORE IN A COIL: When a coil is fitted with a soft-iron core, the coil, when energized, will draw the core inside the coil. The magnetic field strength of a coil can be greatly increased by keeping the iron core in the center of the coil. See Figure 24-9.

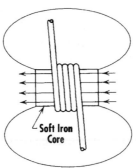

Figure 24-9 Soft-iron core in a coil.

Because soft iron is magnetic and has a low reluctance, more of the flux lines will concentrate in it rather than in the air around it. Remember, the more flux lines present the stronger the field. Soft iron is used as the core material because hard iron would be permanently magnetized, rendering the coil inoperative. This is the principle used in contactors, starters, relays, and solenoid valves used in the air conditioning and refrigeration industry.

Unit 25: Electricity And Magnetism At Work:

Electricity and magnetism are used in almost every component in the electrical circuit of an air conditioning or refrigeration system. Electromagnetism is the energy that causes electric motors to operate. All relays, contactors, and starters, solenoid valves, and some types of thermostats function because of magnetism. It can be seen from this that an air conditioning or refrigeration technician must have a good, solid understanding of basic magnetism.

RELAYS: *Relays* are what is known as electromechanical devices that are used to control electric power to other components in the electrical circuit, such as fan motors, damper motors, and many other such devices. They are equipped with a soft iron movable core called an armature. There is a contact or set of contacts mounted on the *armature* which are called movable contacts. The *movable contacts* match a set of *stationary contacts* to either make or break the electrical circuit as desired for proper system operation. See Figure 25-1.

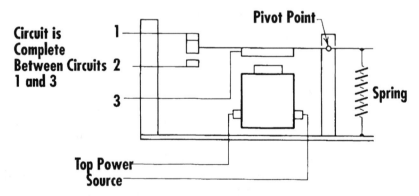

Figure 25-1 Relay cut-away.

Relays may be equipped with a variety of switching actions. The contact may be *normally open* **(NO)** or *normally closed* **(NC)**.

In operation, when electrical power is supplied to the

coil an electromagnetic field is built-up around the coil. Because there is a sift = iron core inside the coil, the core is pulled into the coil, bringing armature along with it. The armature causes the movable contacts to either move away from or move to touch the stationary contacts: thus, either making or breaking the electrical circuit. When the electric current is interrupted to the coil, the armature spring pulls the core out of the coil, taking the armature with it. The contacts move to either make or break the circuit as desired.

CONTACTORS AND STARTERS: A contactor or starter is a device for repeatedly completing or interrupting and electric power circuit. These devices have many features to provide the desired functions. Each has electromagnetically operated contacts that make or break a circuit in which the current exceeds the operating circuit current of the control. Each may be used to make or break voltages which are different from the control circuit voltage. A single starter or contactor can be used to switch more than one circuit. These devices are used for switching heavy current, high voltage, or both.

In operation, the armature is pulled into the coil when current is supplied from the control circuit. Electric current flows through the coil and causes electromagnetism, which attracts the armature. The armature carries a set of movable contacts that match a set of stationary contacts. The electrical circuit to the motor, or other device, is completed through these contacts and the device begins operating. When the controlling device is satisfied, the control circuit is broken. The electromagnetic field around the coil collapses and the armature drops out of the coil, taking the movable contacts with it. The controlled device stops and waits for a signal from the controller for the next operating cycle.

SOLENOID VALVES: These valves are widely used in all types of air conditioning and refrigeration systems. They are used as electrically operated line stop valves and perform the same function as hand operated shutoff valves.

However, because they are electrically operated, they may be conveniently operated in remote locations by any suitable electric control device. The automatic control of refrigerants, brine, or water frequently depends on the use of solenoid valves. These valves are very popular for use when a pump-down cycle of the refrigerant system is desired.

Solenoid valves also operate on the electromagnetic principle. A solenoid valve is a simple form of an electromagnet consisting of a coil of insulated copper wire or other suitable conductor which, when energized by the flow of an electric current, produces an electromagnetic field such as iron and many of its alloys. In this way an armature (frequently called a plunger) can be drawn up into the center of the solenoid coil. See Figure 25-2.

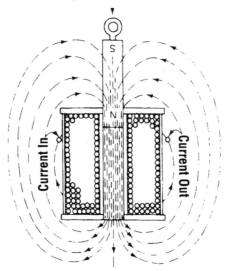

Figure 25-2 Solenoid coil and plunger.

When a stem and pin are attached to this plunger, the valve will open or close a valve port by energizing or deenergizing the solenoid coil. This magnetic principle constitutes the basis of design for all solenoid valves.

THERMOSTATS: Snap-acting thermostats have a movable contact mounted on a bimetal blade and a stationary contact mounted on the base of the thermostat. A permanent magnet is located close to the stationary contact and attracts the bimetal blade and the movable contact after a certain temperature inside the conditioned space has been reached. See Figure 25-3.

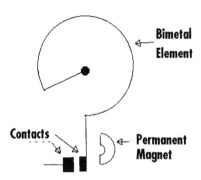

Figure 25-3 Bimetal type thermostat.

The magnetic field caused by the permanent magnet and the spring action of the bimetal blade causes the movable contact to go against the fixed contact with a snap, thus reducing the arcing when the contacts close.

As the temperature inside the space changes (either warmer or cooler), the bimetal blade tends to pull the movable contact away from the fixed contact. Because the movable contact is in the magnetic field, the strength of the bimetal blade is not yet sufficient to overcome the magnetic field created by the permanent magnet. As the temperature of the bimetal blade changes more, it tends to bend. Soon it has enough strength to overcome the magnetic field of the permanent magnet and the movable contact moves away from the stationary contact with a positive snap. This breaks the control circuit and the system responds as it was designed.

ELECTRIC MOTORS: The simplest electric motor possible is made up of a rotor, a permanent magnet mounted on a movable shaft, and two magnetic poles mounted on the outside motor shell. See Figure 25-4.

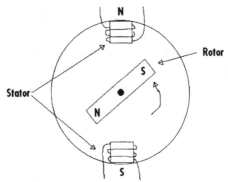

Figure 25-4 Simple electric motor.

The magnetic poles are actually coils of wire; they are stationary poles and are referred to as the *stator*. These coils of wire produce strong magnetic fields when electric current passes through them.

The amount of starting torque built into the motor circuit is the basic difference among the single-phase motors that are used in the air conditioning and refrigeration industry. The method by which torque is obtained is the major difference between these motors.

The magnetic laws of attraction state that like magnetic poles repel and unlike magnetic poles attract. The pole marked with an N (north) in Figure 25-5, is magnetically attracting the S (south) pole of the rotor. Because the stator is stationary, the rotor moves toward the stator poles and rotation of the shaft begins.

ELECTRIC METER: A basic electric meter can be made with a solenoid coil and a soft-iron core to measure the electric current flowing through a conductor. See Figure 25-5.

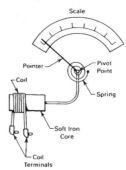

Figure 25-5 Basic electric meter motor.

As current passes through the coil a magnetic field is built up and draws the core into the coil. The strength of the magnetic field is relative to the amount of current flowing through the conductor. Thus, the core is drawn into the coil a distance related to the strength of the magnetic field. The opposite end of the core is fitted with a pointer and spring. The pointer indicates the amount of current flowing through the conductor and the spring tends to hold the core back, preventing its being drawn completely into the coil with a weak magnetic field.

SECTION 2. Single-Phase Electric Motors

INTRODUCTION:

The laws of magnetism and magnetic induction that were discussed in units 21, 23, 24, and 25 are the basis on which all electric motors operate. It should be clear that an understanding of these principles be gained before the basic theory of electric motors is attempted.

Unit 26: Electric Motor Theory:

The most basic electric motor possible consists of a rotor, a permanent magnet mounted on a movable shaft, and two magnetic poles mounted on the outside motor shell. See Figure 26-1.

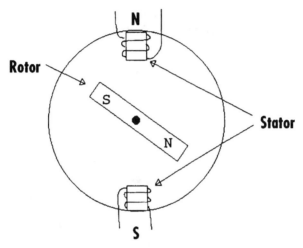

Figure 26-1 Simple electric motor.

These magnetic poles are simply coils of wire. Since they are fastened to the motor shell they are stationary and are called the stator. These poles cause a strong magnetic field when the electric current passes through them.

The basic difference between single-phase motors that are used in air conditioning and refrigeration systems is the amount of starting torque the motor has. The methods used to produce this torque is the difference between them.

The attraction and repulsion caused by magnetic fields are what makes an electric motor operate. The N (north) stator pole in Figure 26-1, is attracting the S (south) pole of the armature because of the attraction of different magnetic fields. The stator is stationary and cannot move, therefore, the rotor, mounted on bearings and designed to move, moves toward the stator pole which starts the shaft to rotating.

The electricity used in the United States is 60-Hz alternating current. Any equipment used on this type of electricity must be designed for that use. Because the current reverses directions 60 times per second the polarity of the stator coils also reverses 60 times per second.

This changing polarity of the stator poles causes a push-pull action on the rotor. The rapidly changing polarity prevents the rotor from stopping when it comes in line with the stator pole with the same polarity. The movement of the rotor causes it to pass the in-line position with the stator coil. See Figure 26-2.

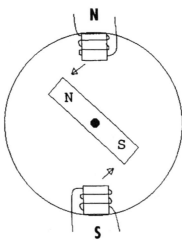

Figure 26-2 Repulsion and attraction.

At this exact time the stator poles change polarity and the N pole becomes the S pole and the S pole becomes the N pole. The S pole of the rotor is repelled by the S pole of the stator and is attracted to the N pole of the stator. Thus, the push-pull action continues until the electricity to the motor is turned off. In theory, the speed of the rotor will automatically turn at 60 revolutions per second or 3600 revolutions per minute. Depending on the number of poles in the stator.

However, the major problem in electric motors is not in running them but in starting them. If the rotor should stop with the S pole of the rotor in line with the S pole of the stator, the motor would not start again. See Figure 26-3.

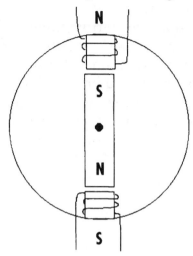

Figure 26-3 Rotor and stator poles in line.

Rotation will not usually begin even when the poles are opposite in polarity because the repelling forces will be at an angle that will prevent the motor from starting.

However, if another stator pole is added the motor can be caused to start turning. See Figure 26-4.

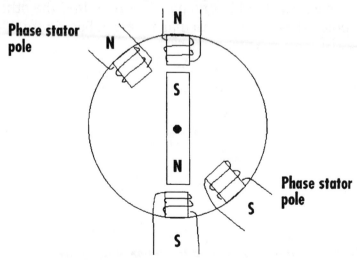

Figure 26-4 Phase stator pole.

The N phase pole on the left will attract the rotor S pole at the same instant that the main stator N pole does. The added pole, known as the start winding, auxiliary winding, or the phase winding, will provide the needed attraction to the S rotor pole to start the motor turning.

However, there remains a problem in starting the motor. Should the rotor stop between the main pole and the phase pole the rotor would not start turning when electrical power is applied to it. See Figure 26-5.

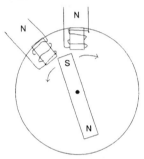

Figure 26-5 Equal rotor pole attraction between the main and the phase poles.

When the polarity of the rotor poles is created, the rotor is pulled equally toward both poles and will not start.

One solution would be for the motor circuit to build a strong magnetic field in one pole before it does the other. As each pole takes on its particular polarity. See figure 26-6.

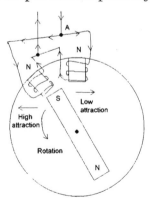

Figure 26-6 High and low magnetic field on stator poles.

Both motor windings are connected to the same power source. The electric current enters the motor circuit at point A. At this point the current is divided. Part of it flows to the left and some of it flows to the right. When the current is caused to flow to the left before it flows to the right, the stator pole on the left will get a strong N polarity before right stator pole does. Because of this the earlier stronger N polarity will attract the rotor to the left to start the motor turning.

If we had two-phase electricity this would not be a problem. But we do not, so we place a capacitor in the circuit to cause the electric motor circuits to think there is two-phase electricity being applied to it.

Unit 27: Split-Phase Motors

Split-phase motors are popular in motor sizes that range from about 1/20 to 1/3 horsepower. They are used in oil burners, blowers, pumps, and fan applications. These types of applications require a motor with moderate starting and running torque.

These motors are of the single-phase induction type that have a nonwound rotor and stator. The windings are placed in insulated slots of the laminated steel core. See Figure 27-1.

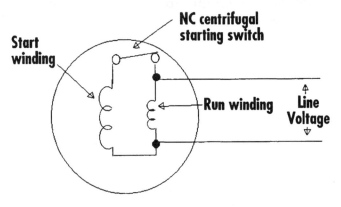

Figure 27-1 Winding placement in a split-phase motor. (The two windings are mixed internally. The smaller of the two wires is the start winding.)

The stator windings are two separate windings. They are known as the run winding, and the phase, start, or auxiliary winding. For these two windings to be effective they are connected in parallel. See figure 27-2.

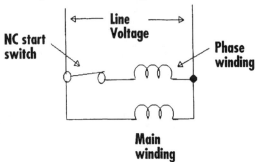

Figure 27-2 Connection of split-phase motor windings.

As can be seen in the drawing, the phase winding is placed so that the magnetic attraction will effectively cause a two-phase condition inside the motor to start it turning and help to bring it up to the normal running speed. When the motor has reached about 75% of its normal running speed a starting switch will open the electric circuit to the phase winding and take it out of the circuit. The split-phase motor now operates as a single-phase induction motor.

Unit 28: Permanent-Split Capacitor (PSC) Motors

PSC motors are popular throughout the air-conditioning and refrigeration industry. Most of the compressors up to five tone capacity use them. They are used on fans, pumps, and blowers just to name some of the uses. PSC motors have a medium starting torque and they operate very efficiently. When they are used for compressor operation, they should be on a system that will have time for the refrigerant pressures to equalize during the off cycle so that the compressor will start in an unloaded condition.

All PSC motors have a running capacitor connected in series with the phase (starting) winding. See Figure 28-1.

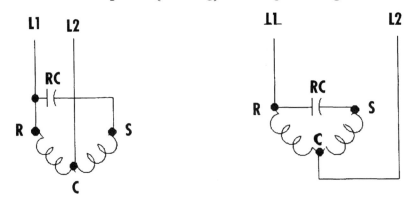

Figure 28-1 Run capacitor connected in the circuit of a PSC motor.

It is connected between the motor run and start terminals. The running capacitor has two important purposes. It causes the split-phase electrical condition required to cause the motor to start running. The capacitor stays in the circuit to cause the phase-shift for efficient motor operation. The two windings, the run and the start, are made from different size wire. The smaller wire is used in the start winding. The starting winding has a higher resistance than the running winding, this helping in keeping the split-phase condition between the two windings. The start winding in PSC motors cannot carry a large current for a very long time. It will quickly burn out and become shorted. Ruining the motor.

Because of this, the running capacitor must have a low Mfd rating. The low rating will reduce the flow of current through the winding to prevent damage to it.

Unit 29: Capacitor-Start/ Capacitor-Run (CSCR) Motors

CSCR motors start with a start and run capacitor, when the motor has reached about 75% of it's normal running speed the start capacitor is taken out of the circuit. An internal starting switch may be used to remove the start capacitor or an external relay may be used for this purpose. These motors have both a high starting torque and the high running efficiency of the PSC motor.

The PSC electrical circuit is the same as that described above for the PSC motor. However, there is additional components and wiring to include the capacitor start condition. The capacitor start circuit includes a start relay, a start capacitor, and the necessary wiring to wire the circuit into the motor circuit. On hermetic type compressors the relay must be used so it is accessible and is not inside the compressor housing where it could possibly cause contaminants and debris inside the refrigerant system. See Figure 29-1.

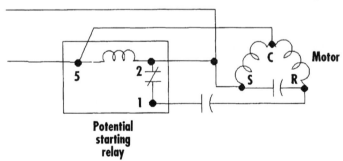

Figure 29-1 Relay connections for CSCR motors.

When the motor is energized, it starts as a CSCR motor. When its speed has reached about 75% of its normal operating speed the start relay is energized from the counter EMF from the start winding and causes its contacts to open. Thus taking the starting components out of the circuit. It then runs as a PSC motor.

Unit 30: The Shaded-Pole Motor

Shaded pole motors have very little starting or running torque. They are very small motors available in wattage ratings. Usually around 1/10 horsepower. They are popular on refrigerator condenser and evaporator fans. Shaded pole motors are also popular on timer motors used for defrost controls and switching timers for turning on and off equipment at predetermined times.

The stator poles are made from laminated steel. In these types of motors the pole pieces are very different from those in other types of motors. Each pole piece has a slot cut in the face of the pole piece. This slot is what is called the shaded pole of the motor. See Figure 30-1.

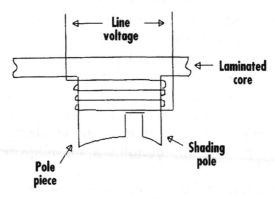

Figure 30-1 The shading pole in a shaded pole motor.

It is a type of induction motor, however, it is wound differently from those of other induction type motors. There is a shading coil placed in the shading slot. The shading coil may be of the wound type or it may be a solid piece of copper placed around the shaded portion of the pole. See figure 30-2.

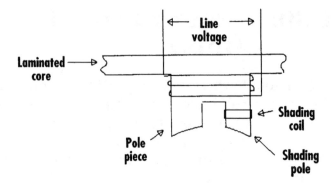

Figure 30-2 Method of shading the pole in a shaded pole motor. (Sometimes the shading is done with fine copper wire and other times a solid copper band is used.)

The loop around the shading pole is not connected to the power source. Also, it must be a closed loop without any type of openings or splits in it. The rest of the pole piece is where the main windings are installed.

When used on ac applications, when power is applied to the main winding, the magnetic field surrounding the coil changes direction constantly. Thus causing the magnetic field to change from N to S. As the ac current flows through the motor windings, the magnetic field is first built to the maximum in one direction and then to the maximum in the other direction. The purpose of the shaded pole band is to slow the speed of the building and collapsing of the magnetic field.

This action creates a condition that when the magnetic field of the main winding is increasing and collapsing at the normal rate, the magnetic field around the shaded pole is changing at a different rate.

Thus, each of the poles peak in magnetic strength at different rates. At some time on the way to a peak, the magnetic field of one of the motor poles will be ahead of the other just long enough to keep the motor turning. At that particular time, the magnetic strength is not equal in each motor winding. A torque is developed and the motor will continue to run. The motor will always turn toward the shaded part of the pole.

Shaded-pole motors are not very efficient. Usually only about 35% efficient. However they are so small that this lower efficiency is not a detriment to using the motor. Usually they run at a higher temperature, therefore, they have a shorter life span than do some of the other types of motors. They should not be used in high temperature applications.

Unit 31: Multi-Speed Motors

Multi-speed motors have been used in the air conditioning industry for several years. Basically, there are two different types of multi-speed motors. **(1)** consequent pole, and **(2)** The permanent-split capacitor (PSC) motor.

There are two ways to change the speed of multi-speed motors. They are to either change the frequency of the electric current to the motor or change the number of stator poles.

The speed of a consequent pole motor can be changed by either putting in or taking out more stator poles. Actually the run winding is divided to change the number of poles that are active in the motor. A two-speed motor run winding is shown in Figure 31-1.

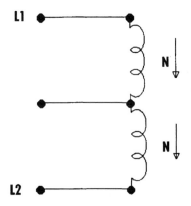

Figure 31-1 A two-speed motor run winding.

When the electricity is connected so that it must pass through the complete winding in only one direction. There is only one magnetic polarity caused in the winding. When the winding is connected to the power as shown in Figure 31-2, the electricity flows in each half of the winding in different directions. This causes the magnetic polarity to be different in each half of the winding. Now there is two magnetic polarities in the run winding. Thus, there are two magnetic poles rather than one as shown in Figure 31-1.

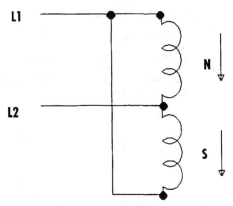

Figure 31-2 Magnetic polarities in multi-speed motors.

Theoretically, a two pole motor will turn at 3600 RPM. The theoretic speed of a four-pole motor is 1800 RPM.

A disadvantage of the consequent pole motor is that it has wide variations in its speeds. When the magnetic poles are changed to four-poles the motor speed changes from 3600 to 1800 RPM. When the consequent pole motor speed is changed it has an increase in its torque. That is it will handle a larger load when in the low speed than it will when it is in the high speed position. Consequent pole motors are generally used for two-speed compressor motors. These motors are not very useful in fans and blowers used in air conditioning applications.

MULTI-SPEED FAN MOTORS: This type of fan motor has been popular in the air conditioning industry for several years. Most of them have from two to five speeds. Their main purpose is to turn fans and squirrel cage blowers. Figure 31-3 shows the wiring for a three- speed motor.

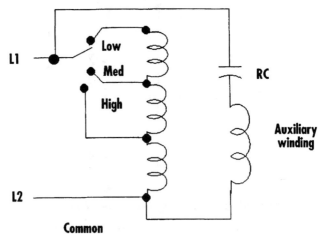

Figure 31-3 Wiring connections for multi-speed motors.

The run winding has been split so that low, medium, and high speeds can be hade with only one motor. The start winding is in electrical parallel to all the run winding, and includes a run capacitor. Notice that the start winding is connected to each side of the applied voltage before it is split in the run winding. The speeds of this motor are possible by placing more resistance in the run winding. Notice that the low-speed uses all three of the run windings. The medium speed uses two thirds of the run winding and high speed uses only one-third of the run winding. Usually a fan relay is used to change the speeds for the needed operation. Usually low- or medium-speed for heating and high-speed for cooling. The purpose of the fan relay is to either place or remove more resistance from the run winding. This resistance causes a decrease in current flowing through the winding and therefore reduces the amount of torque that the motor has when the specific coils are either energized or de-energized.

These motors usually have speeds of 1625, 1500, and 1350 RPM. There is less RPM difference between each of these speeds when compared to the consequent pole motor. Multi-speed motors will generally have from one to four ohms resistance in the run winding and from 10 to 15 ohms in the start, or auxiliary winding. The windings in these motors have a high impedance. This is why these motors can be operated at different speeds without overheating.

These motors are not suitable for use in high torque applications. A complete wiring diagram for these motors is shown in Figure 31-4.

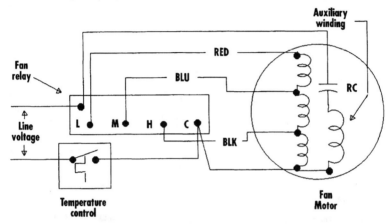

Figure 31-4 A complete wiring diagram for multi-speed motors.

Unit 32: Dual-Voltage Motors

Single-phase motors are sometimes available with more than one voltage rating. The most popular types use 120 or 240 VAC. These motors have two run windings and one start winding. See Figure 32-1.

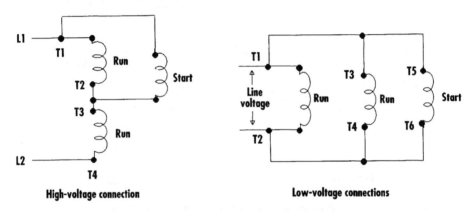

High-voltage connection   Low-voltage connections

Figure 32-1 WIndings and connections of dual-voltage motors.

The run windings are labeled T1 and T2, and T3 and T4. The start winding is connected between T5 and T6. Each of the windings are rated for 120 VAC. In Figure 32-1 the motor has been connected for 240 VAC operation. Note that the two run windings are connected in series to drop the voltage to 120 VAC through each one. The start winding is using 120 VAC because it is connected in parallel with one run winding and in series with the other.

When the motor is to be used on a 120 VAC circuit, the run windings are connected in parallel. See Figure 32-2.

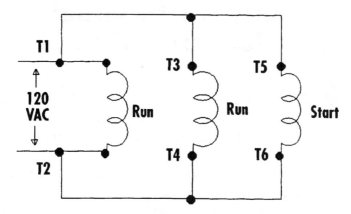

Figure 32-2 Connections for 120 vac dual-voltage motors.

When connected in this way each of the run windings and the start winding have 120 VAC applied to them.

Contrary to popular belief, a motor operating on 240 VAC uses the same amount of electricity, under the same operating conditions, as the same motor operating on 120 VAC. If we calculate the power consumption of each motor it will be the same for both motors. A 240 VAC motor will draw half the amperage of the same motor connected to 120 VAC electricity. 240 VAC is generally used because its lower current rating causes less current draw through the circuit. Thus, reducing the amount of voltage drop on the power supply line. In instances when the motor may be located at some distance away from the electric panel, the voltage drop in the line will usually become a factor in choosing the size wire to be pulled to the motor. Otherwise the motor will draw excessive current and will probably be damaged from overheating.

Unit 33: Terminal Identification

Locating the motor terminals of a single-phase motor is not difficult. All single-phase motors have them and it must be known which is which when connecting compressor motors to prevent damage by reversing the windings. The start winding will not carry enough current to run the motor and it will overheat and be damaged, requiring replacement of the motor or compressor-motor.

The common terminal is where both the start and run windings are connected to one side of the power line. The run winding is connected to the other power line. The start winding is connected through the start components to the same power line as the run terminal. See figure 33-1.

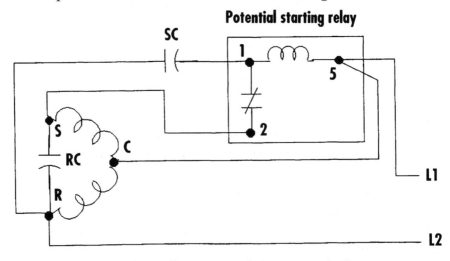

Potential starting relay

Figure 33-1 Connections of common and the motor windings.

An ohmmeter is used to determine the resistance of each of the terminals to another terminal. First set the ohmmeter on Rx1 then zero the ohmmeter so that more accurate readings can be taken. Draw a picture of the exact location of each of the terminals. Now measure the resistance from one terminal to another until all three have been tested from one to another. Be sure to mark on the drawing the resistance reading between each set of terminals as they are made and draw a line between the terminals. See figure 33-2.

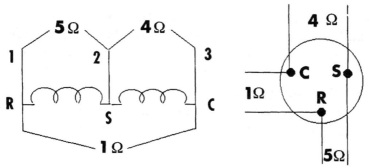

Figure 33-2 Determining start, run, and common compressor motor terminals.

Now use the following rule: The most resistance will be between the run and start terminals; the medium resistance will be between the start and common terminals; the least amount of resistance will be between the run and common terminals.

The most resistance is between terminals 1 and 2; the medium resistance is between terminals 2 and 3; the least resistance is between 1 and 3. From this the common C terminal is terminal number 3, the start **(S)** terminal is number 2, and the run **(R)** terminal is terminal number 1. The resistance between the C terminal to the S terminal should equal the resistance between terminals R and S.

Unit 34: Reversing Motor Rotation

Some motors are designed so that the rotation can be changed without damaging the motor. The major types are the: capacitor-start, run (SCR); capacitor-start, capacitor-run (CSCR); and the permanent-split capacitor (PSC) motors. When a dual-voltage motor is used the type of voltage used should be correct before changing the wiring to change the rotation. The wire can now be changed without trouble. See Figures 34-1 and 34-2.

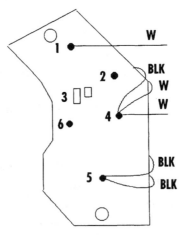

Figure 34-1 Connecting motor for one direction operation. (Be sure to check the motor diagram for the motor being used.)

Some manufacturers include a plug on the outer shell of the motor to make reversing direction easier. The plug can be plugged in for the motor to run in the correct direction.

Those that do not have this reversible plug the wiring must be done inside the terminal enclosure on the motor. When changing the direction of rotation the capacitor is rewired inside the motor terminal enclosure, be sure to follow the motor manufacturer's wiring instructions. Figure 34-3.

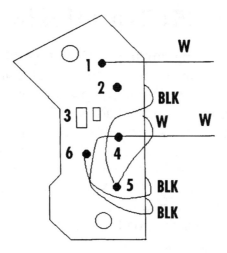

Figure 34-2 Connections for motor in the other direction of rotation. (Be sure to check the motor diagram for the motor being used.)

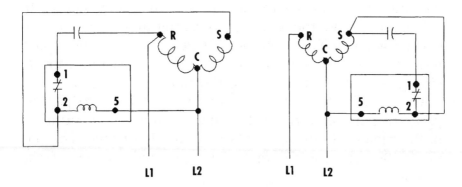

Figure 34-3 Capacitor connections for motor rotation.

In one direction of rotation the capacitor is in electrical series with the run winding. When used in the other direction, the capacitor is connected in electrical series with the start winding.

CAUTION: Do not attempt to change the direction of rotation on hermetic or semi-hermetic compressor motors. This will cause the windings to overheat and be damaged requiring changing the motor-compressor unit.

Unit 35: Troubleshooting

Troubleshooting single-phase motors can sometimes be confusing. But the trained technician should be able to this without undue frustration. The practice of changing a motor because it will not run is a bad practice. However, when the troubleshooting procedures are known, there is little problems in troubleshooting single-phase motors.

Some of the problems will be that the motor will not start, motor starts and then stops, motor current too high, open windings, shorted windings, open internal overload, and open external overload. Each of these will be discussed in the following paragraphs.

MOTOR WILL NOT START: The first thing to check for is to make sure that there is electricity to the motor terminals. To do this remove the cover from the motor terminal box and check the voltage at the terminal connections. If voltage is found the problem is something else. Further checking is needed. When no voltage is found at the motor terminals, the problem is probably a blown fuse or a tripped circuit breaker. Either replace the fuse or reset the circuit breaker. If the fuse blows or the circuit breaker trips, the problem requires further checking. Such as, the continuity of the motor windings, the starting circuit, or other components. Be sure that the motor is connected for the correct voltage.

MOTOR STARTS AND STOPS: When this problem occurs there could be something wrong in the starting circuit, a weak running capacitor, a bad starting relay, a bad overload, or too much current draw.

When the motor starts and then stops, probably the first thing to check is the current draw. If it is too high the motor overload is interrupting the voltage to the motor. If a starting relay is used check to make sure that it is interrupting the starting circuit at the correct time. This can be done by checking the current draw while the motor is starting and after is starts, if possible. If the current does not drop, the starting relay is bad and must be replaced. Be sure to use the correct one to prevent future problems. A weak run capacitor could cause this problem on PSC motors because it is drawing too much current and causes the overload to overheat and stop the motor. To check the run capacitor, replace it with a new one of the correct mfd and voltage rating. If this does not solve the problem the motor must be replaced.

Bad motor or compressor bearings could be causing the problem. If the motor has an external shaft, check to see if it turns easily or is stiff. If it is stiff the bearings are dry and need oiling. If oiling loosens the shaft so that it will turn easily, the bearings will probably cause more problems in the near future. It is usually better to have the bearings replaced or replace the motor to prevent future problems with it. If the motor shaft turns easily check it for sideways movement. If movement is present, the bearings are worn and must be replaced. Either replace the bearings or replace the motor. The cause of the over current condition must be determined and repaired.

MOTOR CURRENT TOO HIGH: Most of the things that will cause a motor to draw too much current are overloaded, bad bearings, or too high or low voltage.

When the motor is overloaded it will usually overheat and eventually one of the windings will burn apart, causing a shorted or open condition. The reason for the overload must be found and repaired or the motor will be ruined. An overloaded motor is usually caused by bad bearings, dirty condenser, or too small a motor has been installed.

Bad bearings will cause an overload on the motor because they do not allow the shaft to move freely. Thus, the motor requires motor current to move the shaft. The motor may not be able to bring the load up to speed when the bearings are bad. If the bearings are found to be bad they must be replaced to remove the overload. When an air cooled condenser is dirty, enough air cannot flow through the coil to condense the refrigerant. The compressor then must pump against a higher pressure which causes an overload on the motor. The condenser must be cleaned before proper operation can be resumed.

When a problem with the applied voltage is found all the components back to the electrical panel must be checked and any problems must be repaired. Be sure to check the contactor contacts if the voltage is too low. If the trouble is all the way back to the electrical panel, the power company must be informed of the problem. Then after the problem has been repaired, all the equipment should be checked for correct current draw. This is to make certain that there was no damage to any of the motors.

Open windings are sometimes the result of overheating. This condition may also be caused by moisture in a refrigeration system and it has caused the motor insulation to deteriorate to the point that the wire shorted out and burnt apart. To check for open windings, set an ohmmeter on Rx1 and zero the meter then check the resistance between each of the pairs of motor terminals. See Figure 35-1.

When a winding is open there will be no continuity

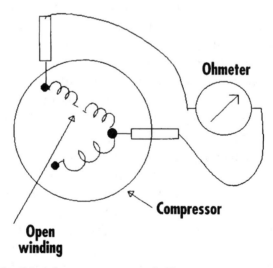

Figure 35-1 *Checking for open motor windings.*

between the two terminals on its ends. When an open is indicated the motor must be either replaced or rewound before the system will operate. If the motor is in a semi- or hermetic compressor, the compressor must be replaced.

A grounded winding is one that has burnt apart and one end of the wire is touching the motor frame. To check for a grounded winding set an ohmmeter on Rx10,000 and zero it. Then check for continuity between the motor terminals and the motor housing. When it is a semi- or hermetic compressor check between each of the terminals and the compressor housing. Any hermetic motor of 1 horsepower or less should have at least 1 million ohms resistance between the windings and the housing. See Figure 35-2.

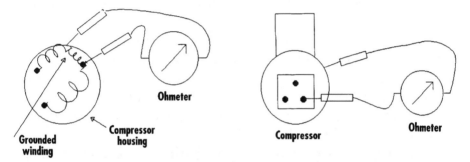

Figure 35-2 Checking for grounded motor windings.

This condition will usually cause the fuses to blow or the circuit breaker to trip. Be sure to check for a bad motor winding when the fuses are blown or the circuit breaker is tripped.

Shorted windings are when the wire used to make the winding has burnt apart and has gotten back together and leaving out some of the winding. The motor will usually draw more current than normal. See Figure 35-3.

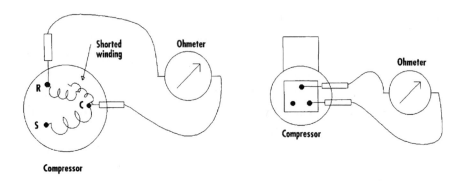

Figure 35-3 Checking for shorted motor windings.

Sometimes the short will be enough to cause the fuses to blow or the circuit breakers to trip.

To check for a shorted winding, set the ohmmeter on Rx1 and zero the meter. Check between each of the pairs of terminals on the motor. The winding that is shorted will indicate less continuity than it should. Sometimes the manufacturer's specifications must be used to determine what the resistance should be.

Testing Motor Winding Insulation: The reason that insulation is used is to keep the electricity from traveling in a path where it is not wanted. The insulation used in electric motors is to cause the electricity to flow through the windings from one end to the other without bypassing any of the winding or flowing to the motor housing where it could be dangerous.

Any insulation will allow some electrical leakage to ground. The windings are exposed to contaminants, rough use, oil, heat, vibration, and expansion and contraction. All of these cause the insulation to start breaking down. The resistance between the motor windings and the housing is very high. Usually an ordinary ohmmeter cannot be used to accurately measure it.

The winding-to-housing resistance is more accurately measured with a megohmmeter. A megohmmeter is very useful in determining the condition of a motor winding. The most useful megohmmeter is the 500-V because it is not strong enough to rupture a good winding and will still give a good accurate reading of the winding under question. A megohmmeter will measure a resistance of from 1 ohm to 1000 megaohms (MΩ).

The megohmmeter can find almost any leakage through the insulation. It can also give an indication of impurities in a refrigeration system, such as; dirt, moisture, or acid because they lower the affect of the insulation. This condition will usually result in a burnt-out hermetic compressor motor.

The megohmmeter has two leads that are used to make the tests. One test lead is placed on a cleaned motor terminal and the other is placed on a cleaned spot on the motor housing. Or they may be placed on pairs of motor terminals to test for open or grounded motor windings. See Figure 35-4.

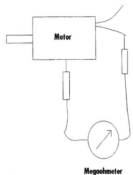

Megaohmmeter

Figure 35-4 Checking motor winding resistance with a megohmmeter.

After connecting the meter leads to the correct place, turn on the megohmmeter. When the meter needle stops rising the maximum resistance has been reached. Remember that each motor will have different readings because of the different conditions that they are used in. Semi- and hermetic compressor motors will usually have about 1000 ohms per volt or one megohm per 1000 volts.

If at all possible, run the motor for about 10 or 15 minutes to warm the components before making the tests. The following are some readings and what they mean to the service technician:

100 MΩ or more indicate that the motor windings are in good condition.

50 to 100 MΩ indicates that there is moisture in the windings, or the refrigeration system

20 to 40 MΩ indicates that there is too much moisture in the refrigeration system or motor windings. If in a refrigeration system; install new refrigerant driers in both the suction and liquid lines. If the motor is used in a high moisture condition consider moving it to a drier place.

Below 20 MΩ indicates that the refrigeration system is severely contaminated and the compressor motor will probably fail very soon. Replace the refrigerant charge and install new large capacity driers in both the liquid and suction lines. If the motor is installed in a high moisture condition, replace it with a moisture proof motor.

Open Internal Overloads: There are two types of internal overloads. The line break type and the thermostatic type. The sensing elements are placed in the windings to measure the temperature at that spot. The line break type will open the common connection inside the compressor housing. It will open the wiring to the common terminal if either an over heated condition or an over current condition exists. This type will open the circuit with a set of contacts located in the common wire inside the compressor housing. See Figure 35-5.

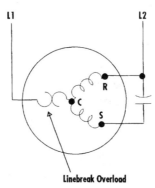

Linebreak Overload

Figure 35-5 Line break overload connections.

When the temperature of the winding reaches break point the contacts will open stopping the compressor. If this overload goes bad the compressor must be replaced. Sometimes the contacts can be made to close by jarring the compressor housing or cooling the compressor housing down with a trickle of water over it. Do not get any water into the terminal compartment to prevent possible shorting of the wiring. Another way is to allow the compressor to cool for several hours and then attempt to start the compressor motor. However, most customers will not allow this procedure. If the contacts will not close, the compressor must be replaced. Be sure to find the cause of the problem before leaving the job or it will only recur at a later date.

Thermostatic Internal Overload: The thermostatic overload has the same purpose as the line break internal overload. However, the thermostatic overload has terminals to the outside of the compressor housing that are connected to the control circuit. See figure 35-6.

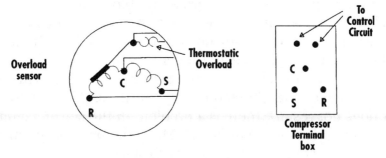

Figure 35-6 Thermostat overload connections.

These external terminals to the control circuit can be temporarily jumped to get the compressor running and cool down the motor, if there is sufficient refrigerant charge in the unit. When using this procedure be sure to check the amperage to make certain that it drops to within the rating of the motor. If the amperage remains higher than the rating stop the compressor and locate the trouble. To continue to run the compressor could cause the motor to burn and increase the cost of repairs. Always check for the cause of the problem before installing a new compressor, or at least before leaving the job. Otherwise, the problem will recur later.

External Overloads: External overloads are located on the outside of the compressor housing. See Figure 35-7.

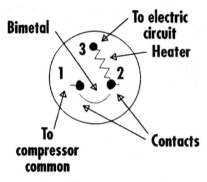

Figure 35-7 External motor overload.

When they are suspected of giving a problem, their terminals can be easily jumped to see if they are open. If they are found open, cool the overload by blowing air over it to cool it sown. When the contacts reset the compressor will restart. If the compressor does not restart when the overload has cooled, jumper the terminals and see if the compressor will run. If it does replace the overload with the correct one so that protection will continue. If the compressor does not restart, there is a problem in the starting circuit or the compressor motor is bad. Make the necessary repairs to get the system operating properly again.

Section 3:
Three-Phase Motors

INTRODUCTION:

Three-phase motors are very rugged, and they are more dependable than the single-phase motors. In most air-conditioning and refrigeration applications the three-phase squirrel cage-induction type is most-popular. There are other types of three-phase motors; such as, the wound rotor and the synchronous type.

Three-phase power and the motors that operate on it are somewhat stronger than the single-phase type discussed earlier. The operating cost is not much different than the single-phase motors, but they can carry heavier loads with less problems. This is because they do not have the extra starting components that single-phase motors must have.

Three-phase electricity has three power legs to the component that it is used to power. Thus, the three-phase power has less displacement between the power legs than the single-phase does. This is what allows the motor to have a higher starting and running torque than its single-phase counterpart.

Unit 36: Three-Phase Motor Principles

All three-phase motors use the rotating magnetic field for operation. Three factors that cause the magnetic field to rotate are:

1. The three voltages change magnetic polarity on a regular basis.

2. The way the stator windings are place around the rotor.

3. Each voltage of three-phase power is 120° out of phase with the others.

The displacement of each of the three phases are 120° out of phase with each other. See Figure-36-1a.

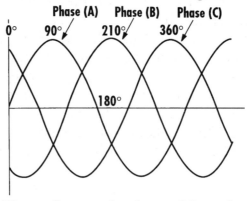

Figure 36-1a Difference between the phases of three phase electricity.

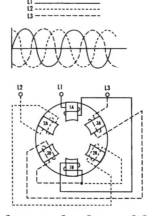

Figure 36-1b Difference between the phases of three phase electricity.

As shown, the stator is wound as a two-pole, three-phase motor. A three-phase motor has two poles per phase. The actual motor will appear somewhat different than the drawing. However, the drawing is used this way to help in understanding how the three-phase motor operates. It should be noted that the pole pieces for each phase are located directly opposite each other. Each pair of the pole pieces are wound and connected to the same phase from the power supply. When wound in this manner each of the pole pieces in the same power leg will have opposite magnetic polarities at the same time.

A more clear presentation of how the magnetic field rotates around the inside of the motor housing. See Figure 36-2.

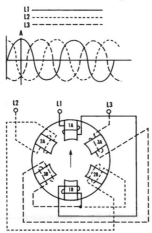

Figure 36-2 Rotating magnetic field of a three-phase motor.

The line labeled 1 has been drawn through the three phases of the power supply. This is an illustration of the phase angle of the three voltage at that instant in time. The arrow that takes place of the rotor shows where the most magnetic strength is at that same time. This also indicates that the arrow is pointing in the "N" (north) direction of the magnetic field. Also, it should be noted that phase 1 is at its positive magnetic strength peak. Also note that phases 2 and 3 have less magnetic strength than phase 1. This causes pole pieces 1A and 1B to have the greatest magnetic strength at this time.

The line labeled 2 in Figure 36-3 shows that the voltage of power leg 3 is zero.

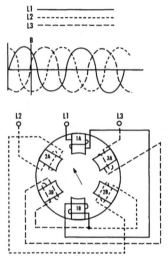

Figure 36-3 Magnetic field strength between pole pieces 1A and 2A, and 1B and 2B.

The voltage of power leg 1 is less than peak positive. The voltage in power leg 2 is less than peak negative. Thus the highest magnetic strength is between the pole pieces 1A, 1B and 2A, 2B.

The line labeled 3 shows the the power leg 2 is at the maximum peak negative. Power legs 1 and 3 are something less than maximum positive. At this point the magnetic field is stronger between pole pieces in phase 2A and 2B and 3A and 3B. See Figure 36-4.

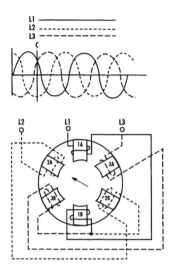

Figure 36-4 Magnetic field strength between pole pieces 2A and 2B.

The line labeled 4 shows that power leg 1 voltage is at zero. Power legs 2 and 3 are something less than maximum and oppose each other. At this time, the magnetic field is stronger between pole pieces 2A and 2B and 3A and 3B. See Figure 36-5.

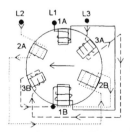

Figure 36-5 Magnetic field strength between pole pieces 3A and 3B, and 2A and 2B.

The line labeled 5 shows that the power leg of phase 3 is at its peak positive and power legs 1 and 2 are less than maximum in the opposing directions. At this point the magnetic field is strongest between pole pieces 3A and 3B. See Figure 36-6.

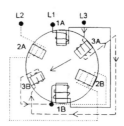

Figure 36-6 Magnetic field strength between pole pieces 3A and 3B.

The line labeled 6 shows that phase 2 has zero voltage. Power leg 3 is less than peak maximum. Power leg 1 is less than peak negative. The magnetic field is the strongest between the pole pieces 1A, 1B, and 3A, 3B. See Figure 36-7.

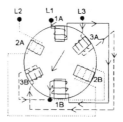

Figure 36-7 Magnetic field strength between pole pieces 3B, 1B and 3A, 1A.

The line labeled 7 shows that power leg 1 is at the maximum negative. Power legs 2 and 3 are at some point less than maximum and are opposing each other. The magnetic field is again strongest between pole pieces 1A and 1B. However, the magnetic polarity of the two pole pieces is opposite than when it first started. The magnetic polarity has changed because the current is changed in the stator winding. See Figure 36-8.

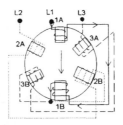

Figure 36-8 Magnetic field strength between pole pieces 1A and 1B.

The line labeled 8 Shows that power phase 2 is at the maximum positive. Phases 1 and 3 are something less than negative maximum. The magnetic field is strongest between pole pieces 2A and 2B. See Figure 36-9.

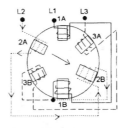

Figure 36-9 Magnetic field strength between pole pieces 2A and 2B.

The line labeled 9 shows that power phase 3 is maximum negative at this time. Power legs 1 and 2 are something less than maximum positive. The magnetic field is now the strongest between pole pieces 3A and 3B. See figure 36-10.

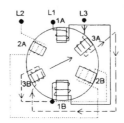

Figure 36-10 Magnetic field strength between pole pieces 3A and 3B.

The line labeled 10 shows that power phase 1 is at the peak positive. Phases 2 and 3 are less than negative maximum. Now the magnetic field is strongest between pole pieces 1A and 1B. We have now completed one complete cycle of the three-phase power system. The magnetic field has moved through the complete 360° around the stator winding of the motor. See Figure 36-11.

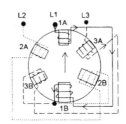

Fig. 36-11. Magnetic field strength between pole pieces 1A & 1B. This is a complete cycle of three-phase voltage.

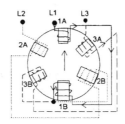

**Figure 36-11 *Magnetic field strength between pole pieces 1A and 1B*
*This is a complete cycle of three-phase voltage.***

Any time that two of the power lines of a three-phase power system are connected to different legs than those shown here, the magnetic field will be in the opposite direction. The motor will turn in the opposite direction. To change the direction of rotation of a three-phase motor switch the connections of any two power legs and the motor connections.

THREE-PHASE MOTOR CONNECTIONS: There are two ways to connect the stator windings of a three-phase motor to the source of power. Either the Y connection or the Delta type connection is used to complete this process.

No matter which one is used, the windings are connected so that only three stator leads come out of the motor. Therefore, connecting the motor to the power line is simple.

Dual voltage, three-phase motors have stator leads that are numbered.

The most common voltages used in air conditioning and refrigeration work are 240 and 480 VAC. The stator wiring diagrams are sometimes labeled "high" and "low" voltage. For connecting the motor to operate on 240 (low) VAC, use the diagram shown in Figure 36-12.

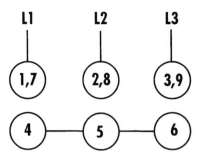

Figure 36-12 Stator connections for 240 (low) vac power. Terminals 4, 5, and 6 are connected together but not to a power lead.

When the motor is to be used on 480 (high) VAC the stator wires should be connected as shown in Figure 36-13.

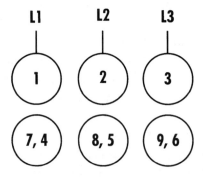

Figure 36-13 Stator connections for 480 (high) vac power. Terminals 7 and 4 are connected together, 8 and 5 are connected together but not to a power source.

Unit 37: The Squirrel Cage Motor

This motor is named because of the shape of its rotor. They are made with bars placed around the outside of the rotor. They are imbedded in the soft iron laminated core of the rotor. These bars are connected at each end by a metal ring. The shape then looks like a squirrel cage.

In operation this type of motor operates as an induction type motor. Induction means that the current flowing through the rotor is caused by a voltage induced by the magnetic field of the stator. When the stator is energized, the rotor is not moving. The magnetic field generated inside the stator windings cut across the bars in the rotor. If the motor is rated for 1800 rpm; three factors that control the speed are:

1. The number of turns of wire that are cut by the magnetic in the rotor. In the case of a squirrel cage motor it is the number of bars in the rotor that are cut rather than wire.

2. The speed of the magnetic field that is cutting the wires or bars.

3. The strength of the magnetic field. A stronger magnetic field will cause more current to be induced into the rotor.

When the power is first applied to the motor, the rotor is not moving. At this time the maximum amount of current is induced into the rotor. This induced voltage causes a current to flow through the rotor. As this current flows through the rotor, a magnetic field is generated around each of the rotor bars. This causes the magnetic field in the rotor to be attracted to the magnetic field in the stator. As the magnetic fields become more attracted to each other the motor starts turning in the same direction that the magnetic field is turning. This gives the motor the torque necessary to start the motor turning.

When the motor starts turning faster, the rotating magnetic field cuts across the bars in the rotor slower than when the rotor was not turning. Because the rotor is now turning the magnetic field cuts across the rotor bars at a slower speed. This causes less current to be induced in the rotor. As the current flow in the rotor is lowered, the current flow in the stator coil is also lowered. We can see form this that the most current flows through both the stator and rotor when the motor is not turning.

For this reason AC induction motors require more starting current than running current.

Torque: The torque produced by an alternating current induction motor is determined by three factors, as follows;

1. The stator magnetic field strength

2. The rotor magnetic field strength

3. The difference in the phase angle of the current in the stator and the rotor.

It is impossible for an induction motor to run at synchronous speed. As the rotor approaches the speed of the rotating field in the stator there are fewer magnetic lines of flux to cut the bars in the squirrel cage rotor. Thus, the current flow through the rotor is less. When there is no current flow in the rotor, there could be no magnetic field in the rotor. Because of this there would be no torque produced. This is one of the factors listed above that decides the amount of torque that is produced in an induction type motor.

Note, that when an induction motor is not connected to a load, it will turn close to the synchronous speed. The speed of an ACV induction motor is controlled by the amount of torque needed. When operating at no load conditions, the motor will produce only the amount of toque needed to overcome its own friction. Therefore, when there is no load on the motor it will turn at a speed close to the synchronous speed of the rotating magnetic field in the stator.

When a load is connected to a motor, more torque is needed to turn the load at the required speed. When the load is placed on the motor shaft, the rotor slows. This causes more of the magnetic lines of force to cut across the rotor bars at a much faster speed. This causes more voltage to be induced into the rotor causing more current to flow. The increase in current causes a stronger magnetic field in the rotor. The increased magnetic field causes more current to flow through the rotor and produce more torque at the rotor shaft. When this happens more current will flow to the motor.

The amount of torque produced in an induction type motor is the amount of phase angle between the current in the stator and the phase angle of the current in the rotor. When the phase angle between the current in the rotor and the phase angle in the stator is 90° out of phase the greatest amount of torque is produced. The current flow through and AC circuit having only resistance is in phase with the voltage. The current flow through a circuit that has only induction, the current and the voltage are 90° out of phase. With the current lagging the voltage.

Squirrel Cage Motor Testing: A three-phase squirrel cage motor can be successfully tested using an ohmmeter. The stator winding can be tested for open shorted or grounded windings. When testing the stator winding for an open condition, test the resistance of each winding in turn. Be sure that the electricity to the motor is off and the wires to the motor leads are disconnected before making any of these tests. See Figure 37-1.

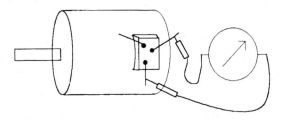

Figure 37-1 Testing stator windings of a three-phase motor.

Before testing, set the ohmmeter on R x 1 and zero the meter. The resistance of all three should be the same. If there is no resistance indicated for any winding, the motor has an open winding and must be replaced or repaired.

When testing for grounded windings, check between each of the motor leads to the motor housing. See figure 37-2.

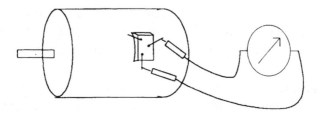

Figure 37-2 Testing for grounded stator windings of a three-phase motor.

Before testing clean the motor housing where the test lead is to be connected so that a good clean connection can be made. Also, zero the ohmmeter before testing the motor. There should not be a reading between any of the motor leads to the motor housing during this test. If there is, the motor is grounded and must be repaired or replaced. It is usually best to use a megohmmeter for this test because the resistance of the insulation on larger horsepower motors is low enough that a reading could be indicated when actually there is no ground present.

A shorted winding is much more difficult to find. A meter that measures the actual inductance of the winding is used. When one winding indicates a lower inductance than the others, it is shorted. The motor must be repaired or replaced.

Another way to check for a shorted three-phase motor winding is, if the motor will run, start it and check the amperage draw of each leg to the motor leads. If one has a higher amperage reading than the others, turn off the electricity and change that lead with another of the leads to the power line. Be sure to remember which ones were changed so that the rest of the test will be valid. Then start the motor and check the current draw again. If the leg drawing the highest current did not change, the motor has a shorted winding. It must be either repaired or replaced. If the leg with the highest current draw moved with the power supply lead, the trouble is in the power supply to the motor. The problem must be found and repaired or the motor will eventually fail.

Unit 38: The Wound Rotor Motor

Wound rotor motors are generally used to operate large air conditioning or refrigeration systems. The wound rotor motor stator winding is the same as the squirrel cage motor. However, the rotor does not have the bars placed in it. The rotor of the wound rotor motor, as its name implies, has a wire wound rotor. It has coils of wire much like the stator windings. The number of coils in the stator will be the same as those in the rotor. When the windings of the rotor are connected together they form a "Y" connection. The other end of the rotor windings are connected to a slip ring that is placed on the rotor shaft.

The slip rings are used to connect external resistances to the rotor windings. See Figure 38-1.

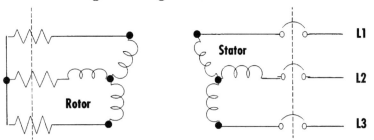

Figure 38-1 Rotor schematic of a three-phase wound rotor motor.

Here the connections for both the stator and rotor connections are shown. This diagram also shows the "Y" connected stator winding being connected to the source of power. The "Y" rotor connections are connected to the three variable resistors. These resisters are mechanically connected together. Thus, when the resistance of one is changed the others are changed the same exact amount.

Carbon brushes are used to connect the external resistances to the slip rings. See Figure 38-2.

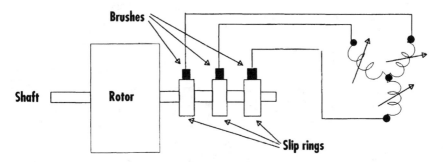

Figure 38-2 Rotor wiring connections for wound rotor motor.

Because the resistance is external of the rotor, it can be changed when needed. This adjustment allows the amount of current flow through the rotor to be adjusted. When the amount of current flow through the rotor is at some point, the same amount of current will flow through the stator.

Some of the advantages of the wound rotor motor are: It limits the amount of inrush current when the motor is in the starting mode. This lower inrush of current eliminates the need for using reduced voltage starters or "Y" or delta starting of the motor. They have a high starting torque. The resistors used limit the amount of current flowing through the rotor windings which tends to keep the phase angle between the stator and the rotor current very close to 90°.

Slip Ring Motor Starting: When the wound rotor motor is first started is started, it has the maximum amount of resistance added to the rotor circuit. As the motor gains speed the resistance is taken out of the stator circuit. This removal of resistance continues until all the rotor windings are shorted together.

There are many different methods used to take the resistance out of the rotor circuit during the starting of the wound rotor motor, such as: the current limit control, rotor speed control, slip frequency control, definite time control.

In the current limit control method, the amount of current flowing through the stator windings is limited. The rotor speed type control is also known as the slip frequency control. The definite time control method uses relays to limit the amount of resistance in the rotor circuit. See figure 38-3.

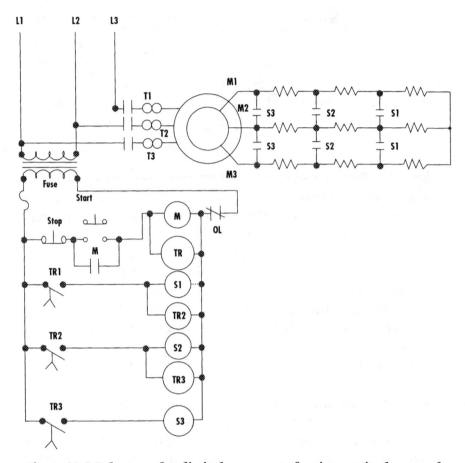

Figure 38-3 *Relays used to limit the amount of resistance in the wound rotor circuit.*

This diagram shows the motor circuit above. In the bottom part of the circuit, a control transformer is used to step down the line voltage to the amount used in the control circuit.

The operation is:

1. When the power to the motor is turned on, a circuit is completed through "M"the motor starting coil, TR1 coil, and to the overload contacts. As the "M" coil is energized, all the "M" contacts close. The three load contacts at the top of the diagram close to energize the stator winding. The "M" contact, below the start button, is known as the holding contact. It is used to maintain the circuit to the "M" coil when the start button is released. In this position the maximum resistance is in the rotor circuit and the motor will start in its slowest speed.

2. The "TR1" is a timer relay. We will assume that all the timers used are set for a three second time delay. When the "TR1" relay is energized it begins the timed function. When three seconds have passed, the "TR1" contacts close. This makes the circuit to the "S1" and the "TR1" coils.

3. When the coil "S1" is energized, both sets of the "S1" contacts are closed to short out the last three resistors in the rotor circuit. When all the resistance is removed the motor speeds up to the next higher speed. When coil "TR2" is energized, its timing sequence starts.

4. When the three second timer period has passed, contacts "TR2" close to complete the circuit to coil "S2" and "TR3".

5. When the "S2" coil is energized, both sets of "S2" contacts close to remove the next set of resistors from the circuit. This allows the motor to speed up to its next higher speed. When coil "TR3" is energized, its timing sequence begins.

6. When three seconds have passed, the contact "TR3" closes and completes the circuit to "S3." When coil "S3" is energized it closes both sets of "S3" contacts to remove the the last set of resistors from the circuit. The motor now speeds up to its highest speed.

7. When the stop button is pressed, the circuit to coils "M" and "TR1" are opened. Coil "M" then opens all the "M" contacts. The stator winding is disconnected to the source of power. Coil "TR1" opens the contact "TR1" immediately. This deenergizes coils "S1" and "TR2." When coil "S1" is deenergized, both sets of "S1" contacts open. When the coil "TR2" is deenergized contacts "TR2" opens. The opening of contacts "TR2", breaks the circuits to coils "S2" and "TR3." When coil "S2" is deenergized, both sets of "S2" contacts open. Then contact "TR3" opens when coil "TR3" is deenergized. Coil "S2" is also deenergized to open both sets of "S2" contacts.

8. If any of the safety devices open, it will shut the system down in the same sequence that pressing the stop button would.

Wound Rotor Motor Testing: Testing the stator of a wound rotor motor is just the same as testing the stator windings of a squirrel cage motor. However, the rotor winding is tested very much like testing the stator winding.

The rotor can be tested for an open in the winding by using an ohmmeter set to RX1 and zeroing the meter before starting the test. Next check the continuity between each of the slip rings. See figure 38-4.

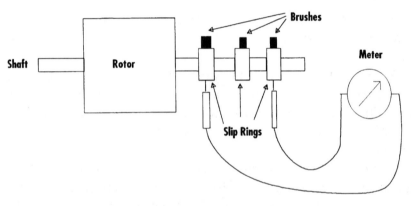

Figure 38-4 Checking continuity between slip rings.

The resistance should be the same between each pair of the slip rings.

The rotor can be tested for a ground by using the ohmmeter set to RX1 and the ohmmeter zeroed before the test is started. Touch one of the ohmmeter leads to the motor shaft and the other meter lead to each of the slip rings in turn. See Figure 38-5.

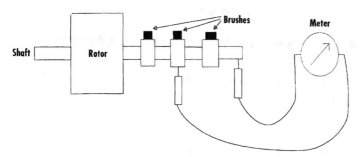

Figure 38-5 Checking continuity between the rotor windings and the shaft.

There should be no continuity indicated between the rotor windings and the ground.

A shorted rotor winding is usually checked with an induction meter. If any of the windings has less inductance than the others, it is shorted. Another way to test for a shorted rotor winding is to start the motor, if it will run, and measure the current draw of the three power legs to the motor. If one of the legs draws more current than the others, change that motor lead to another power lead and check the current draw. If the high current moves to the new leg, there is a problem in the power source. It must be found and repaired or the motor will burn out. If the high current draw remains on the same motor lead the rotor is shorted. The motor must either be repaired or replaced.

Unit 39:
The Synchronous Motor

The synchronous type motor has some characteristics that no other type motor has. Some of these characteristic are:

1. The synchronous motor will run at the same speed whether it has no load or is fully loaded.

2. The synchronous motor is not an induction type motor. It does not need the induced voltage from the stator to cause a magnetic field in the rotor.

3. The synchronous type motor not only will correct its own power factor, it can also correct the power factor to all the other motors that are connected to the same power line.

The stator and rotor are both wound much like the wound rotor motor. However, the rotor is wound with only one continuous set of coils. The synchronous type motor has only two slip rings rather than the three used in the wound rotor motor. The rotor has bars much like those used in the squirrel cage motor. This set of bars are used to start the synchronous motor.

Synchronous Motor Starting: When power is first applied to the stator, the rotating magnetic field cuts through the rotor bars. The cutting action of the magnetic field on the bars induces a current into the rotor bars. The current flowing through the winding causes a magnetic field in the rotor that is attracted to the rotating magnetic field of the stator. This magnetic field causes the motor to start turning in the direction of magnetic rotation in the stator winding. When the rotor has reached the speed of the synchronous field speed, direct current is connected to the rotor through the slip rings on the rotor shaft. See Figure 39-1.

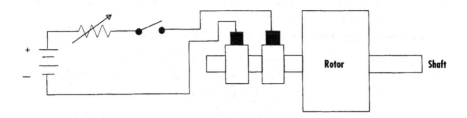

Figure 39-1 Connections for direct current to the rotor through the slip rings and motor shaft.

The application of the direct current to the rotor winding causes the rotor windings to become electromagnets. The electronic field in the rotor locks in step with the rotating magnetic field in the stator winding. Now the rotor turns at the same speed as the rotating magnetic field in the stator windings. When the rotor and the magnetic field in the stator windings are turning at the synchronous speed, there is no more cutting of the lines between the two magnetic fields. This causes the current flow in the rotor winding to stop.

The synchronous type motor has good starting torque and low starting current draw.

The Field Discharge Resistor: When power is first applied to the synchronous type motor , the magnetic field cuts through the rotor winding and bars very fast. This cutting action causes a large voltage to be induced into the rotor winding. To limit the amount of this voltage, a field discharge resistor is placed across the rotor windings. See Figure 39-2.

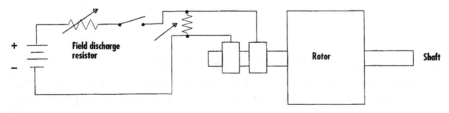

Figure 39-2 Connections for field discharge resistor.

This field discharge resistor also helps in reducing the amount of voltage that is induced into the rotor winding by collapsing the magnetic field when the DC current is disconnected from the rotor rings.

Power Factor Correction: The synchronous motor power factor can be changed by adjusting the amount of DC current to the motor rotor. When the DC current is adjusted so that it is in phase with the voltage, the motor will have a power factor of 100%. This condition is usually considered to be normal excitation for this type motor.

The DC Power Supply: The supply of DC power to the rotor rings can be provided by several means. The most common is a small DC generator installed on the motor shaft. Also, an electronic inverter can be used to change the AC to DC voltage.

Synchronous Motor Testing: The stator winding in a synchronous type motor is tested just the same as any other motor. The rotor winding can be tested using an ohmmeter set on RX1 and zeroed before testing.

To test the rotor for an open winding touch one of the ohmmeter leads to each of the slip rings on the rotor shaft. See Figure 39-3.

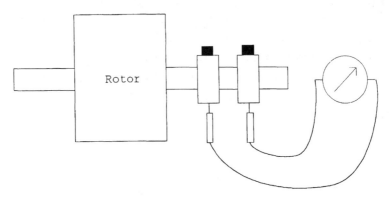

Figure 39-3 Testing the rotor for an open circuit.

Because the synchronous speed motor rotor is designed to be used on DC current, the wire resistance will be high when compared to the resistance wire resistance of a wound rotor motor.

Test the rotor using an ohmmeter set on RX1 and zeroed

before making any tests.

To test for a grounded winding touch one of the meter leads to one of the slip rings. The touch the other lead to the motor shaft. See Figure 39-4.

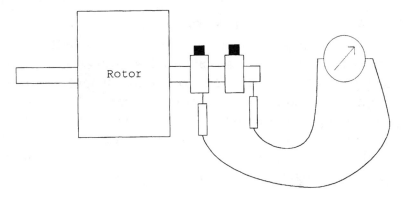

Figure 39-4 Testing the rotor for a grounded winding.

There should be no continuity indicated. If there is the rotor windings are grounded. The rotor must either be rewound or replaced.

Section 4:
Control Devices

INTRODUCTION:

The controls used in modern air conditioning and refrigeration systems are devices that provide protection and or cause the motor or component to operate. These controls may be manually operated, electrically operated, electronically operated, pressure operated, or temperature operated. The technician must have a solid understanding of the controls and the systems in which they function to be proficient in troubleshooting and repair of these systems.

UNIT 40: Disconnect Switch

INTRODUCTION: Disconnect switches are usually of the two or three pole single throw type. A pole is provided for each electrical phase that is to be interrupted. These are blade type poles which are operated manually by changing the position of a lever located on the outside of the box.

Their main purpose is to provide a means of interrupting the electrical circuit when repairs or testing are required. It is a good idea to either place a sign on the handle indicating that there is work being done on the electric circuit and not to turn the switch on, or physically lock the switch in the off position to prevent the switch being turned causing possible personal injury to the technician or damage to the unit.

UNIT 41:
Fused Disconnect Switch

Fused disconnect switches are sometimes used in place of regular disconnect type switches. Their purposes are to provide a manual disconnect in the circuit to be used for testing and repair procedures, and protection for the circuit during operation. They include a fuse to protect each electric circuit and its components. A fuse is used for each phase of the circuit. See Figure 41-1.

Figure 41-1 Fused disconnect switch.

The fuses must be properly sized to provide the required protection. It is a good idea to either place a sign on the handle indicating that there is work being done on the electric circuit and not to turn the switch on, or physically lock the switch in the off position to prevent turning it on and cause possible personal injury to the technician or damage to the unit.

UNIT 42:
Circuit Breaker Panels

Circuit breakers are used to provide overload protection to an electric circuit and its components. These breakers are placed inside a circuit breaker panel. They must be properly sized to provide the required protection for the circuit. Usually there are several circuits located in a single panel, thus, it is a good idea to label each breaker indicating the circuit it protects.

Circuit breakers are not designed to provide repeated manual circuit interruption. Their purpose is to provide circuit protection in case of overload. However, there are times when it is necessary to manually open the circuit with the circuit breaker. It is a good idea to either place a sign on the breaker handle indicating that there is work being done on the electric circuit and not to turn the breaker on, or physically lock the panel door closed to prevent turning the breaker on causing possible personal injury to the technician or damage to the unit.

UNIT 43: Single-Phase Motor Protectors

Most single-phase motor circuits include a temperature actuated set of contacts to protect the motor against overheating, overcurrent, or both. These overload devices may be located inside the motor or placed on the outside of the motor housing. The equipment design will dictate where the overload is located.

INTERNAL OVERLOAD: Internal motor overloads may be the line break type or the thermostatic type. Both types are located precisely in the center of the heat-sink of the motor windings. They are used to protect the motor from both overheating and overcurrent conditions. See Figure 43-1.

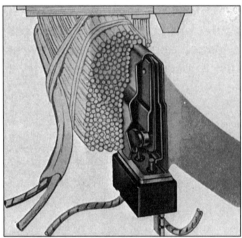

Figure 43-1 Location of compressor-motor internal overload. (Courtesy of Tecumseh products company.)

Line Break Overload: This type of overload is placed in the wire between the common terminal and the motor winding. See Figure 43-2.

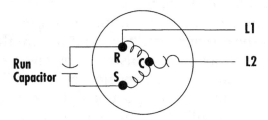

Run Capacitor

L1

L2

Figure 43-2 Line break internal overload diagram.

When either an overheated condition or an overcurrent condition exists the overload contacts will open and interrupt the electrical circuit through the motor. These controls are usually designed to be used in the line voltage circuit.

Thermostat Overload: The internal thermostatic overload protector is also mounted in the motor windings. The contacts may interrupt either the line voltage to the motor, as in the line break, or they may interrupt the control circuit to the motor contactor or starter. This type of overload is sometimes referred to as a pilot duty overload. See Figure 43-3.

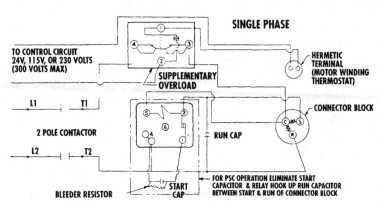

Figure 43-3 Pilot duty overload diagram.
(Courtesy of Tecumseh Products Company.)

The connections between the overload and the control circuit are usually made in the compressor terminal box.

Troubleshooting: To test the internal line break overload, first make certain that the contactor is supplying line voltage to the compressor. Turn off the electric power at the disconnect switch. Remove the wire from the common terminal of the motor. Place the voltmeter between the motor terminal and the disconnected wire and restore the electric power. Be sure to follow the instructions given earlier for using the voltmeter. If there is no voltage indicated the internal line break overload is open. It must be cooled before its contacts will close and complete the circuit. If voltage is found the overload contacts are closed and are not the problem, or the overload contacts reset during the testing procedure. Check to see if the motor will run at this time.

Troubleshooting: To test the internal pilot duty overload, make certain that the control circuit is supplying the correct voltage to the overload terminals. If the correct voltage is indicated, turn off the power to the control circuit. Remove a wire from one terminal to the overload and test between the two terminals with an ohmmeter. Be sure to use the instructions given previously concerning use of the ohmmeter. If continuity is indicated the overload is not the problem. If no continuity is indicated the overload contacts are open. The motor housing must be cooled to the required temperature so that the contacts will close. It is possible to run the motor when this overload is open. Simply connect the two wires from the control circuit together and turn on the unit. Check the operation of the motor to determine what caused the overload to open. Correct the problem and see if the overload has reset. If not the compressor must be cooled or replaced. Do not leave the unit operating with this overload bypassed because permanent damage could possibly occur. If the motor is in a hermetic system, the clean-up would be much more expensive than a simple compressor-motor replacement.

When an internal overload contacts open, the complete motor winding must cool down to the reset temperature of the control. When it has been determined that this is the problem, simply allow cool water to trickle over the housing. Be sure to prevent the water from contacting any of the electrical wiring or connections. It is usually a good practice to turn off the electricity to the unit during this procedure. If this is not possible allow the unit to stand idle for several hours to cool to the required temperature. When it is determined that the overload is defective the complete motor or the motor-compressor must be replaced because there is no way of replacing only the overload.

EXTERNAL OVERLOAD: External overload protectors are placed on the outside of the compressor-motor housing. When found to be faulty they may be replaced. Be sure to use an exact replacement and place it exactly where the old one was and in the same manner. Otherwise the desired protection may not be provided. Usually they are mounted in complete contact with the compressor housing so that the temperature can be properly sensed. See Figure 43-4.

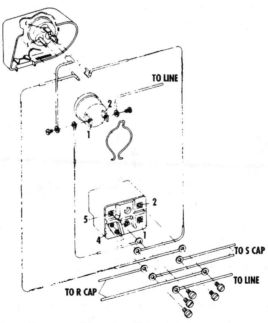

Figure 43-4 Location of external overload.
(Courtesy of Tecumseh Products Company.)

External overloads may be of the line break type or of the thermostat type. See Figure 43-5.

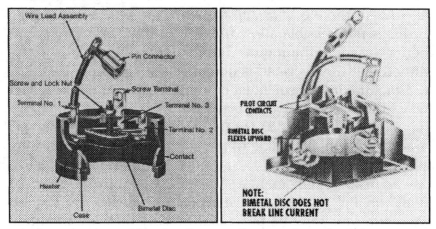

Figure 43-5 External line break (a) and external overload thermostats (b). (Courtesy of Tecumseh Products Company.)

The line break type will have two or three external electrical connections and the thermostat type will usually have three electrical connections. See Figure 43-6.

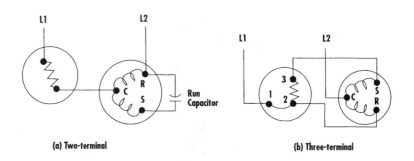

(a) Two-terminal

(b) Three-terminal

Figure 43-6 Overload connections (a) line break, (b) three-terminal.

When a three wire line break overload is used, protection for both the starting and running phase are protected. See Figure 43-6b.

The external type overloads are operated by a bimetal disk. See figure 43-7.

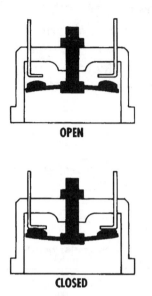

OPEN

CLOSED

Figure 43-7 External overload contact operation. (Courtesy of Tecumseh Products Company.)

The common terminal of the overload is connected into the electric circuit to the common terminal of the motor winding. See Figure 43-8.

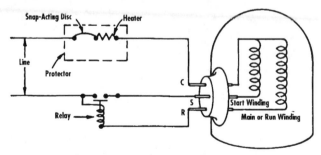

Figure 43-8 External line break overload connections. (Courtesy of Tecumseh Products Company.)

One of the other terminals is connected to the run terminal and the other to the start terminal on the motor. When the resistance heater has provided sufficient heat, indicating an overload condition, the contacts will open and interrupt the electrical circuit to the motor. The motor will be inoperative until it has cooled to the required temperature for the contacts to reset.

The thermostat external overload is mounted in exactly the same way as the line break type. However, the contacts do not open the main electrical circuit to the common terminal of the motor. See Figure 43-9.

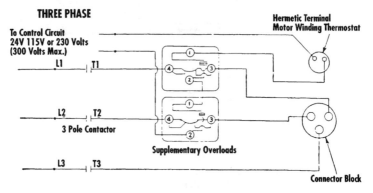

Figure 43-9 External thermostat overload connections.
(Courtesy of Tecumseh Products Company.)

Instead, the control or pilot circuit is opened, which causes the contactor or starter to interrupt the electrical supply to the motor.

UNIT 44: Solid State Protectors

Many applications of solid state controls are used as single-function type controls in the air conditioning and refrigeration industry. These types of controls are used to either protect or control the operation of a piece of equipment. They are used for heat pump system defrost controls, for time-delay to prevent compressor motors from short cycling, for motor protection, and as starting relays used on single-phase motors. They are used for many other types of control functions which will be discussed later in the solid-state section.

TIME DELAY CONTROL FOR SHORT-CYCLE PROTECTION: Heat is the greatest enemy of electric motor winding insulation. If a motor is overheated from any cause for a number of times, failure of the insulation eventually occurs. Overload is a major cause of motor overheating and can be due to low voltage, single phasing, unbalanced voltage, or just plain mechanical overload caused by short cycling of a motor trying to start against a heavy load.

There are time delay control models available for the protection of motors and compressors from damage due to low voltage, short cycling caused by the thermostat or power failure, phase loss, and phase out-of-sequence. See Figure 44-1.

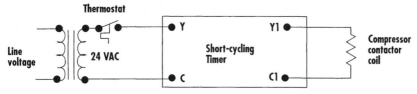

Figure 44-1 *Time delay control for short-cycle operation and wiring diagram. (Courtesy of Watsco, Inc.)*

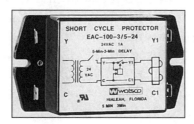

Delay-on-Break: These are solid-state timers that prevent short cycling of compressor motors by delaying a restart due to momentary power failure or thermostat interruption. The time delay begins when the power is interrupted or the thermostat contacts open.

Delay-on-Make: Some of these timers provide a 3-minute time delay occurring on power resumption, providing a random restart of the equipment after a control interruption or a power failure. These types usually have a set delay for both the on and break operations of the control.

ADJUSTABLE SOLID-STATE TIMERS: These types of timers may be adjusted to control a desired delayed on or delayed off function. See Figure 44-2.

Figure 44-2 Adjustable solid state timer.
(Courtesy of Watsco, Inc.)

They are used for delayed on, delayed off, or staggering the starting of heavy equipment to prevent circuit overload. Adjustable timers can be adjusted over a wide range of on and off cycles. They are available for either line voltage or control voltage applications.

Troubleshooting: Troubleshooting solid-state timers is not very complicated. Make certain that there is voltage supplied to the timer and that the load terminals are energized when they are supposed to be. This requires only the AC voltmeter to make these tests. If the timer is not supplying voltage to the load terminals at the correct time replace the timer.

SOLID-STATE STARTING RELAY: Solid-state starting relays use a self-regulating conductive ceramic which increases in electrical resistance as the motor starts, thus quickly reducing the starting winding current flow to a milliampere level. The relay contacts open in about 1/3 second after being energized. This function allows this type of relay to be used on multiple sized motor-compressors without the need of being tailored for each individual size within the specialized current limitations of the relay. They are designed for push-on installation on the motor-compressor-pin connector.

This type of relay is connected into the motor circuit with the ceramic between the line to the motor and the starting terminal of the compressor motor. See Figure 44-3.

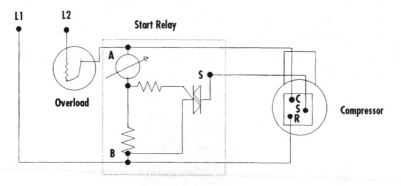

Figure 44-3 Solid-state starting relay connections.

This places the ceramic in electrical series with the starting winding. When the power is supplied to the relay, the ceramic material begins to heat up which turns off the relay (opens the relay contacts) in about 1/3 second. When this happens, the current flow through the start winding is reduced to the milliampere level. When the electricity is interrupted to the motor circuit the ceramic material cools to start the next cycle. These types of starting relays are not suitable for use on short-cycling applications or for capacitor-start motors.

These devices are available for larger motors and have

wiring connections so that they can be installed in the correct circuit to the motor. They operate in exactly the same manner as described above.

Troubleshooting: To check the operation of these controls it is necessary to determine if the relay is energizing the winding on start and if it drops out of the circuit after a brief operating period. This can be done by placing the ammeter on the wire to the starting terminal. In cases when it is impossible to check the amperage at this point measure the total line amperage to the motor-compressor. If the amperage fails to drop or if the starting circuit is not being energized, (both indicated by excessive current draw) replace the relay. Be sure to use a proper replacement.

SOLID-STATE HARD-START KIT: Solid-state hard-start kits are used to provide additional starting torque when PSC motor starting problems occur. The relay uses a ceramic material that has a positive temperature coefficient. This means that the material increases in resistance as its temperature increases. See Figure 44-4.

Figure 44-4 Solid-state hard-start kit. (Courtesy of Watsco, Inc.)

The purpose of the hard-start kit is to provide a surge of electric current that will last long enough to start the motor. This additional current then drops to a much lower value. The motor is then operated in the PSC mode. When the electric circuit is completed to a PSC motor, current flows through the start winding and through the parallel combination of the run capacitor and the low-resistance ceramic material. See Figure 44-5.

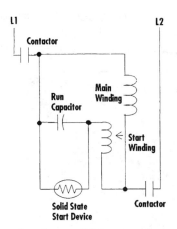

Figure 44-5 Solid-state hard-start kit connections.

The low resistance of the ceramic during the start mode allows more current to the start winding, and also reduces the angular displacement between the main winding and the start winding during this period.

While this surge of current increases the motor starting torque, it is also flowing through the ceramic and causing it to heat to its high resistance state. This is ideal for starting PSC motors because the ceramic will be at its lowest resistance state only long enough to overcome the initial inertia of the compressor. When the ceramic is at its high-resistance state there is only a very small amount of current flowing through the start winding.

To determine if the hard-start kit is functioning properly it is only necessary to determine if the ceramic is energizing and dropping out the start winding. This can be easily accomplished by measuring the amperage to the motor. If the motor starts and the amperage remains high, the kit is not dropping out the start winding. If the amperage is high and the motor does not start, the ceramic is not completing the circuit to the start winding. In either case the kit must be replaced to correct the starting problem.

SOLID-STATE DEFROST CONTROL: All heat pump systems use some type of defrost control to remove the frost and ice from the outdoor coil during the heating season.

Manufacturers use various types of systems to provide this part of the cycle. However, the solid-state defrost control is the most popular on newer units. See Figure 44-6.

Figure 44-6 Solid-state defrost control and wiring diagram.
(Courtesy of International Controls and Measurements.)

The purpose of removing the ice and frost is to keep the unit operating as efficiently as possibly. The solid-state module uses one thermistor to initiate the defrost cycle and one to terminate it. The initiating transistor is attached to the outdoor coil at a point usually specified by the equipment manufacturer. The termination transistor is placed to sense the air flow through the outdoor coil. There is usually a pressure switch used to provide a safety and to terminate the defrost cycle should the module fail to function properly.

The wiring schematic of one type of defrost control is shown in Figure 44-6. There is a DC volt defrost relay, the two thermistors, and a 24-volt connection to the control voltage. Each of these devices have two connections that must be properly made for the control to function as designed. When the temperature of the outdoor coil drops to 30°F the initiation thermistor energizes the defrost relay coil. At this point, 18 volts DC are directed to the defrost relay coil. When the defrost relay is energized the system is changed to the defrost cycle.

After the outdoor coil has been defrosted the air sensing

thermistor senses the air temperature rise across it. When this air temperature has risen to the cut-in temperature the system is switched back into the heating mode.

When the control is suspected of giving trouble, the technician must first determine if it is functioning as designed. About the only tools needed are an accurate thermometer, an Ac voltmeter, a DC voltmeter, and an ohmmeter.

Troubleshooting: Check the outdoor coil temperature and the air temperature across the coil. If they are above the cut-in setting of the module, the system should not go into the defrost cycle.

If the system should not go into the defrost mode at a temperature higher than 30°F. If it does either the initiation thermistor or the module is bad. To check the resistance of the transistors, use the manufacturer's specifications for that module. When the proper resistance is not indicated the module is bad. Another way to check the thermistors is to check to make certain that 18 volts DC are at the defrost relay terminals when the outdoor coil temperature is low enough to demand defrost. If 18 volts DC are present at the defrost relay terminals and the system does not go into defrost the module is bad and must be replaced. If there is not 18 volts DC detected at the defrost relay terminals and the system is in the defrost mode the defrost relay is stuck in defrost. The next step is to make the indicated repairs and place the system back in operation.

UNIT 45: Magnetic Overload

The magnetic, or Heinemann, overload is used to protect larger electric motors. They are available in either single units or as overload devices used as an integral component of the starter. See Figure 45-1.

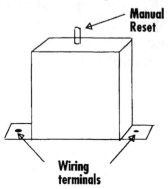

Figure 45-1 Magnetic overload.

These are pilot duty devices that interrupt the control circuit which in turn interrupts the line voltage to the motor. See Figure 45-2.

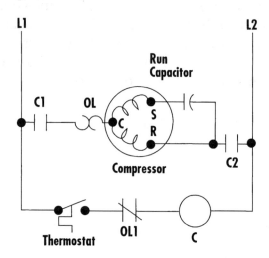

Figure 45-2 Schematic of magnetic overload in circuit.

These devices consist of a magnetic coil connected to a movable core placed inside a tube filled with the correct amount of either oil or silicone to provide a time delay function, a set of line voltage contacts, and four terminals. In operation, the current flowing to the motor is directed through the magnetic coil surrounding the oil, or silicone, filled tube. The coil wire is sized according to the current draw of the motor. When the motor current increases the current flowing through the coil increases the magnetic field around the tube. When this magnetic field reaches the designed strength the core allows the oil or silicone to flow through a small hole in it and cause the pilot duty contacts to open. When they open the voltage to the control circuit is interrupted and causes the starter to drop out and the motor stops.

Troubleshooting: Magnetic overloads are easy to troubleshoot because they have only a set of contacts and a magnetic coil. About the only tools needed are an ohmmeter and an ammeter. Turn off the electric power to the unit and remove the wire connections at the overload. Check the continuity of the pilot duty contacts in the control circuit and then the continuity of the magnetic coil. The manufacturer's specifications may be required to determine the correct resistance of the coil. If continuity is detected at both places, reconnect the wiring and start the unit. Check the amperage draw of the motor and if it is within the motor nameplate rating and the overload still trips off, the overload is defective. Replace the overload with the correct replacement. If an incorrect replacement is used the motor may not be properly protected.

UNIT 46: Relays

These are electromechanical devices placed in an electric circuit to control the power to other system components; such as electric motors, damper motors, strip heaters, and many other components. Relays consist of a movable soft-iron core called an armature. At least one set of contacts and usually several sets are used to control the various circuits used. They consist of both movable and stationary contacts. The movable contacts are placed on the movable armature and the stationary contacts are mounted on the relay frame. One movable and one stationary contact completes a set of contacts. See Figure 46-1.

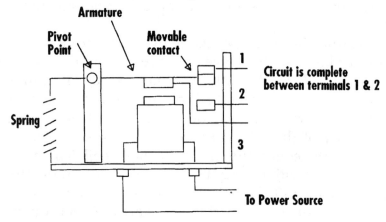

Figure 46-1 Relay components.

Since relays are used for a variety of functions the contacts may be either normally open **(NO)** or normally closed **(NC)**, depending on the requirement. The word normally indicates the condition of the contacts when there is no electricity applied to the coil. There is also an electromagnetic coil placed around the movable armature. The purpose of the coil is to provide a magnetic field to pull in the armature and switch the contacts.

In operation, when electrical power is supplied to the coil an electric field is generated around the movable armature. The soft-iron core is then pulled into the center of the coil, causing the movable armature to follow it. The movement of the armature causes the movable contacts to move to either make a circuit or break a circuit. When the electric current to the coil is interrupted, the armature spring pulls the core out of the center of the coil taking the movable armature along with it. The contacts take their normal position of either normally open or normally closed as required.

Troubleshooting: About the only tools required to troubleshoot a relay is a voltmeter. Apply the proper voltage to the coil and listen to the relay. If there is a clicking noise heard when energized the coil and armature are working. Then check the voltage at the contact terminals to determine if the contacts are opening and closing as desired. If the voltage at the terminals does not change to show contact operation, the contacts are stuck. Replace the relay. If there is no click when the coil is energized then either the coil is open, there is no electricity to the coil, or the armature is stuck. Check to see if electricity is supplied to the coil, if so and the relay still does not operate replace it.

UNIT 47: Starting Relays

Starting relays are used on practically all single-phase motors with the exception of the permanent split-capacitor (PSC) type. The purpose of the starting relay is to remove the starting capacitor and any other starting components that may be used to aid the motor in starting from the circuit.

All electric motors require a higher amperage during the starting period than they do while running. This is because all the components are starting from a stand-still position. Usually the starting current is about six times the normal running current. As the motor gains speed the components are moving faster and requiring less current. When the current draw reaches a specified amount the starting relay removes the starting components from the circuit which also reduces the amount of current flow to the motor.

In operation, when the motor reaches approximately 75% of its normal operating speed, the starting relay is designed to remove the starting components from the starting circuit. Open type motors used in applications requiring a higher starting torque and a capacitor, use a centrifugal switch to remove the starting components from the circuit.

There are basically four types of starting relays used in the refrigeration and air conditioning industry. They are: the **(1)** amperage (current), **(2)** hot wire relay, **(3)** potential, and **(4)** solid state relay. Usually the horsepower and the motor design dictate which type is used.

AMPERAGE (CURRENT) STARTING RELAY: The amperage starting relay is normally used on units having 1/2 horsepower and smaller. See Figure 47-1.

Figure 47-1 Amperage (current) relay. (Photo by Billy C. Langley.)

The relay coil is wired in series with the running winding of the motor. All of the current flowing to the run winding passes through the starting relay coil. Thus, the wire for the coil must be large enough to carry the amperage draw of

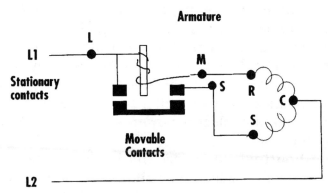

Figure 47-2 Amperage relay wiring diagram.

the motor without burning-up. See Figure 47-2.

The relay contacts are normally open **(NO)**. The armature of the relay which is connected to the contacts is placed inside the relay coil. When the heavy inrush of current to the motor during the start phase builds a magnetic field around the armature, causing the relay to "pull-in" closing its contacts. Current then flows through the contacts to the start winding to provide the extra torque needed to start the motor and bring it up to speed. These are positional type relays. This is, they must be installed so that the weighted contacts and armature will open the contacts by gravity force when the starting current drops to a predetermined value.

Operation: When the thermostat or other controller demands operation of the system, the electrical circuit is completed which supplies current to the motor. The current flowing through the relay coil causes the armature to be drawn into the center of the relay coil, closing the starting contacts. See Figure 47-2. The closed contacts complete the circuit to the starting winding of the motor, providing the required shift in the electrical power to start the motor. As the motor approaches about 75% of its normal running speed the counter EMF, which is 90 degrees out of phase with the applied voltage, builds up in the start winding and reducing the current flow through the winding. The reduced current allows the magnetic field in the coil of the starting relay to be weaker so that the gravity force on the relay armature and contacts can open the contacts interrupting the current flow to the start winding of the motor. The motor remains in the running mode until the operating controller demands that the unit stop. The relay contacts will remain open until electric power is again applied to the motor windings.

Troubleshooting: Amperage relays must always be sized to the motor on which they are used. They are usually sized according to the amperage draw of the motor. A relay that is rated for a larger motor may not cause the contacts to close because it is the current draw that causes the contacts to close. When this occurs, the starting circuit is left out of the starting phase of the motor. It may not start or it may have a difficult time in starting. A relay that is rated to small for the application may not allow the contacts to open, keeping the starting circuit in operation longer than necessary or continuously. This would cause the motor to draw high current for an extended period which may cause damage to the start winding. If a starting capacitor is used, the capacitor may be damaged by staying in the circuit too long. Always use the proper motor protector with a current starting relay.

Hot Wire Relay: The hot wire relay is another form of the current relay. However, it is not the actual current that causes the relay contacts to function as desired, it is the heat generated by the flow of current through a small heater inside the relay. The current flows through a resistance (hot) wire which heats up because of the resistance and provides heat to a bimetal strip that causes the contacts to either open or close. See Figure 47-3.

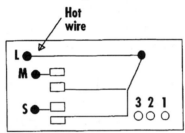

Figure 47-3 Hot wire relay.

This relay is equipped with two sets of contacts, one for the running phase and one for the starting phase. These contacts are normally closed **(NC)**.

Operation: When the thermostat or other controller demands operation of the equipment, the electrical circuit is completed allowing current to flow through both the starting and running contacts in the relay. See Figure 47-4.

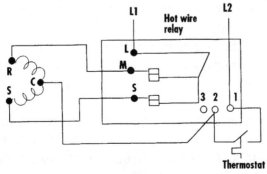

Figure 47-4 Schematic of a hot wire relay.

The current that flows through the resistance wire to the main, or running, winding of the motor causes the wire to heat. In turn this heat causes the bimetal strip to warp opening the starting contacts. The starting contacts are opened and the running contacts remain closed. This is the normal operating position of the hot wire relay. If an excessive amount of current continues to flow through the resistance wire the bimetal strip will warp further and open the running contacts to stop the motor. Both sets of contacts will remain open until the bimetal strip has cooled enough to allow both sets of contacts to close at the same time to start the motor again. The sequence will be repeated until the motor is operating or the relay gets so hot it will not close for a long time.

Troubleshooting: Hot wire relays are nonpositional. They must be sized according to the motor size. When the relay is sized for a larger motor, the starting contacts will remain closed longer than necessary to properly start the motor. Also, if the relay is sized large enough the starting contacts will not open at all. When they are sized for a smaller motor, the relay may stop the motor as though it had an overload. The motor will stop before the controller is satisfied. An overload is not required with the hot wire relay.

Potential (Voltage) Starting Relay: These types of relays operate because of a magnetic field that is generated by voltage flowing through an electric coil wound around a soft-iron core. Potential relays are used on almost all sizes of electric motors. The starting contacts are normally closed **(NC)**. They are opened with a plunger that is "pulled in" (pulled into the relay coil) when sufficient magnetic field is generated.

The relay consists of a set of normally open contacts, an electromagnetic coil that is made from very fine wire, and a base and terminals. There are three necessary terminals on the outside of the relay cover. They are numbered 1, 2, and 5. The other terminals numbered 4 and 6 are sometimes used as auxiliary terminals. See Figure 47-5.

Figure 47-5 Potential; starting relay. (Courtesy of Watsco, Inc.)

Potential relays are nonpositional, but they must be installed where there is a minimum amount of vibration. Vibration will cause the contacts to arc excessively and become pitted and burned quicker than normal. Terminal 5 is connected to both the electrical line to the motor and to the line leading to the common terminal of the motor. Terminal 2 is connected to the start terminal of the motor. Terminal 1 is connected to one of the start capacitor terminals. The other capacitor terminal is connected to the other side of the supply voltage line. See Figure 47-6.

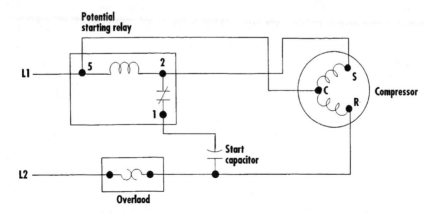

Figure 47-6 Schematic for a potential relay.

Operation: When the thermostat or other controller demands system operation the electrical circuit to the motor is completed. Power is supplied directly to the running winding and to the starting winding through the starting relay contacts between terminals 1 and 2. When the motor reaches approximately 75% of its normal operating speed, the counter EMF increases in the start winding to a predetermined value to "pick up" the relay, opening the contacts and removing the start components from the circuit. The relay is now in its normal operating position. It should be remembered that it is the voltage in the start winding that causes the relay to "pick up" and remove the starting components from the circuit. When the thermostat or other controller is satisfied the electrical circuit to the motor is interrupted and the motor stops. When the motor slows down the counter EMF in the start winding also decreases. When the "drop out" voltage of the relay is reached, the relay armature drops out of the coil and causes the relay contacts to close. The relay is now ready to provide the required starting torque to the motor when demanded.

Troubleshooting: When the motor starts and runs for a short period of time and then stops, the relay should be checked. A simple way to check a potential starting relay for this problem is to remove the wire to terminal 2 on the relay. Make the wire contact the terminal and start the motor. Then immediately remove the wire from the terminal. If the motor continues to operate and draw the correct amount of amperage, the relay contacts are stuck closed or the relay coil is burned out. In either case the relay must be replaced. If the motor tries to start, but just hums and then stops on the overload, either the start capacitor or the relay may be faulty. To check the relay, place a jumper between terminals 1 and 2 on the start relay. Start the motor. If the motor starts remove the jumper immediately. If the motor continues to run and draw the correct amount of amperage, the relay contacts are stuck open. The relay must be replaced. If the motor still just hums and does not start, place a wire on each terminal of a start capacitor of approximately the correct size. Touch one of the wires to terminal 1 and the other wire to terminal 2 on the starting relay. Start the motor. When the motor starts remove the wires from the relay. If the motor continues to run drawing the correct amount of current, the capacitor is bad. The capacitor must be replaced with a correct replacement. It is usually preferred to replace both the starting relay and the capacitor when the capacitor is bad because the relay contacts may be sticking to cause the capacitor to fail.

If the proper relay is not known, a good way to determine the "pick up" voltage is to start the motor and check the voltage between the start and common terminals. Multiply this reading by 0.75. This will indicate the required "pick up" voltage of the new relay.

Solid State Relay: Solid-state advancement has produced a current-sensitive relay that uses a PTC (positive temperature coefficient) ceramic. Current to the starting winding flows through the ceramic and causes it to heat. As its temperature is increased its resistance also increases. These relays are being used to replace a number of other type starting relays because they do not require that a specific relay be used, just that it will operate in the range of horsepower it is designed for. See Figure 47-7.

Figure 47-7 Solid-State starting relay. (Courtesy of Watsco, Inc.)

These relays are available with a built-in starting capacitor, and motor overload.

There is also another type of solid-state starting relay, the universal starting relay, is being used to replace the potential relay. See Figure 47-8.

Figure 47-8 Universal starting relay.
(Courtesy of Sealed Unit Parts Co. Inc.)

This relay operates on time rather than either voltage or current flow. Because it operates on time this relay can be used to replace many other starting relays without regard to accuracy in sizing. To replace a potential relay with a universal relay use the wiring diagram shown in Figure 47-9.

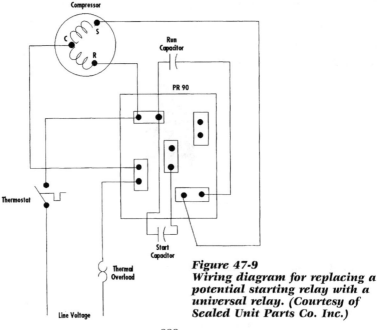

Figure 47-9
Wiring diagram for replacing a
potential starting relay with a
universal relay. (Courtesy of
Sealed Unit Parts Co. Inc.)

To replace an amperage relay with the universal relay use the wiring diagram shown in Figure 47-10.

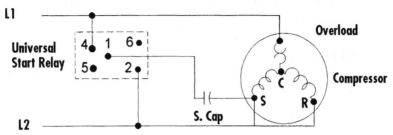

Figure 47-10 *Wiring diagram for replacing an amperage relay with a universal relay.*

Operation: Solid-state relays are wired in series with the start winding of the motor. See Figure 47-11.

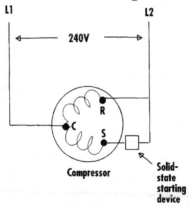

Figure 47-11 *Diagram for a solid state starting relay on a split-phase motor.*

When at ambient temperature they have a very small amount of resistance. When the thermostat or other controller demands system operation electric power goes directly to the run winding of the motor and through the start relay to the start winding. The current flowing through the solid-state starting relay causes it to heat very rapidly. As the resistance is increased the current flow to the start winding is reduced. Quickly taking the start winding out of the circuit. When the thermostat or controller is satisfied the system is stopped. The solid-state starting relay must cool to the ambient temperature before the motor can be restarted because the current cannot flow through the PTC material to provide the necessary starting torque. Therefore, they are not recommended for use on applications where rapid cycling may occur.

Solid-state starting relays can be used in combination with starting capacitors for applications requiring higher starting torque. The capacitor may be either an integral part of the relay or it may be separate. See Figure 47-12.

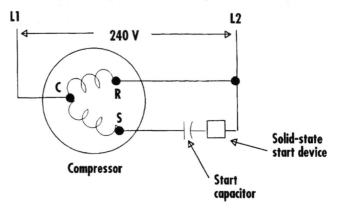

Figure 47-12 Diagram for a solid-state starting relay and capacitor-start motor.

A motor overload protector must be used with these types of starting relays. Some types are available that incorporate a relay, capacitor, and an overload.

Troubleshooting: Troubleshooting the solid-state starting relay is relatively simple. Usually the outside of the relay will show that overheating has occurred. When this happens, the relay is usually defective and must be replaced. Resistance of the relay can also be checked with an ohmmeter to verify that it is bad. Before checking the resistance make certain that the relay is cool, otherwise a faulty reading will be indicated. A low resistance indicates that the relay is good. When a high resistance is indicated the relay is bad and must be replaced. To make certain that the relay is removing the start winding from the circuit place an ammeter on the wire to the start winding. The current flow should be reduced to practically nothing in a very short period of time. If the relay is the push-on type connecting it directly on the compressor terminals about the only way to make this test is to measure the total amperage to the unit to see if it drops very quickly after the motor has started. If the current draw in either test does not drop the relay is defective and must be replaced.

Unit 48: Thermostats

Thermostats are used as primary controls in temperature control applications. They are temperature-sensitive switches that control the operation of the equipment. They are available in many styles, shapes, sizes, and purposes, depending on the system requirements and the type of operation desired. Thermostats have only one function in a control system and that is to sense the temperature inside the structure whether it is in a commercial building, a residence, a domestic refrigerator, or a commercial food case and signal the equipment when to start and stop. They may have fixed control points or multiple variable adjustments. Some types are designed for use in 24 volt control systems and others are designed for operation in line voltage applications.

Cooling thermostats make the circuit on a rise in temperature inside the conditioned space and break the circuit on a fall in the temperature. Heating thermostats make the control circuit on a fall in the temperature of the conditioned space and break the circuit on a rise in the temperature. There are thermostats that are used for both cooling and heating applications. Usually they include a system switch that will allow the selection of heating, cooling, or continuous or intermittent fan operation, or automatic change-over operation.

CONTROLLING ELEMENTS: Basically there are two types of controlling elements used in thermostats. These are the bimetal and the gas-filled bulb. The controlling element is that part of the thermostat that changes position when a temperature change is sensed.

Bimetal Thermostats: The bimetal thermostat is probably the most popular. It uses a bimetal strip to sense the temperature inside the conditioned space. See Figure 48-1.

Figure 48-1 Bimetal in a thermostat. (Photo by Billy C. Langley.)

A bimetal is two pieces of different types of metal hermetically welded together. The two pieces of metal have different expansion and contraction rates which causes the bimetal to either bend more or tend to straighten out on a change in temperature. One end of the bimetal is attached to the thermostat base and the other end is free to move with a temperature change. See Figure 48-2.

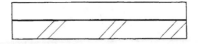

Figure 48-2 A bimetal element.

In most cases they are wound into a helical shape. One end is fastened to the thermostat base and a movable contact is fastened to the other end. A stationary contact is placed on the thermostat base. There is a small permanent magnet placed close to the movable contact which causes a snap action to prevent arcing of the contacts and to make a positive connection between the two contacts. This snap action also eliminates problems of poor temperature control by causing the contacts to remain either closed or open until sufficient force is presented by the bimetal to actuate the movable contact. See Figure 48-3.

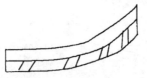

Figure 48-3 Bimetal operation.

This snap action may also be achieved by using a mercury bulb contact on the bimetal. The weight of the mercury inside the bulb will change positions inside the bulb with a temperature change in the conditioned space. There is a ridge placed in the center of the bulb so that the mercury will collect in a puddle before completely changing ends of the bulb. When the temperature change is sufficient to cause the bulb to "dump" the mercury suddenly moves to the other end and either making or breaking an electrical circuit through the bulb. See Figure 48-4.

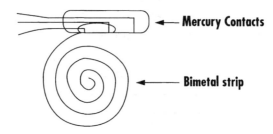

Figure 48-4 Mercury-type thermostat contacts.

Remote Bulb Thermostat: The remote bulb type thermostat uses a gas in a hermetically sealed system to provide the contact movement in response to temperature changes required to make the switch close or open on demand. See Figure 48-5.

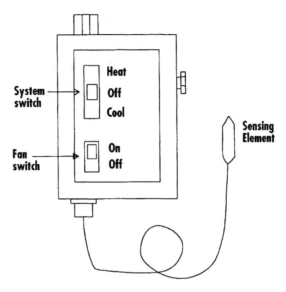

Figure 48-5 Remote bulb-type thermostat.

As the temperature inside the conditioned space changes these changes are transmitted to the thermostat through the diaphragm. When the temperature changes the remote bulb senses the change and either exerts more pressure or less pressure on the thermostat causing the contacts to either open or close, depending on the design of the thermostat. This type of thermostat is used in just about all temperature control applications. The bulb may be installed in and air conditioning duct to sense the temperature there, or it may be installed on a tube to sense the temperature of a coil or some fluid inside the coil.

These types of thermostats do not necessarily have an external bulb attached to them. This type uses the same principle of operation but the bellows is located inside the thermostat and operates without the external bulb.

Line Voltage Thermostat: These types of thermostats are designed to directly control the voltage supplied to the unit. That voltage may be either 120 or 240 volts. See Figure 48-6.

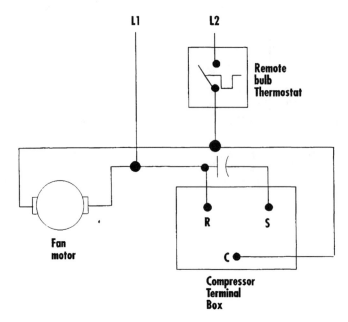

Figure 48-6 Diagram of a line voltage system with a remote bulb thermostat.

They do not respond well to low voltage applications. In most installations these thermostats simply open or close a set of contacts. They are most popular in commercial refrigeration and self-contained air conditioning applications, window air conditioners and commercial refrigeration cases. Typical wiring connections are shown in Figure 48-6.

Low-Voltage Thermostat: Low-voltage thermostats are used on applications where 24 volts are used for the control circuit. This typically includes residential air conditioning, commercial air conditioning, and heating installations. They are available for heating only, cooling only, heating and cooling systems, automatic and manual fan operation, and automatic changeover from cooling to heating or heating to cooling. These thermostats cannot be used on line voltage installations because they are not designed to carry the voltage or the current used in these applications. A typical residential heating and cooling system wiring diagram is shown in Figure 48-7.

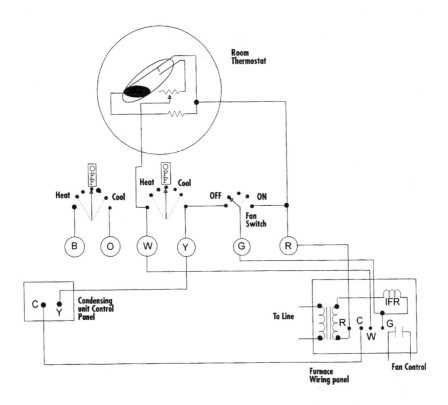

Figure 48-7 *Typical residential low-voltage heating and cooling diagram.*

Low-voltage thermostats are safer than line voltage thermostats. They are cheaper, smaller, more accurate, require smaller wiring, and usually do not require a licensed electrician to wire the thermostat.

Anticipators: An anticipator is a resistor which is placed in the control circuit inside the thermostat. These resistors play a very important part in accurate temperature control of air conditioning systems. Without them the temperature would have such wide variation that it would be almost impossible to maintain any degree of comfort inside the conditioned space. Two types of anticipators are used in comfort conditioning thermostats: heating and cooling.

Heating Anticipator: An unanticipated thermostat will allow the space temperature to have such wide swings that it would be almost impossible to maintain satisfactory conditions. When the thermostat senses a need for heat, the contacts close the circuit to the gas valve. Since no heat is available for the space until the heat exchanger has heated sufficiently, the space temperature may drop another two or three degrees before warm air is actually distributed to the space. This is called system lag. When the temperature has reached the cut-off setting for the space its contacts will open the circuit to the gas valve which will stop the heating process. However, the furnace heat exchanger must still be cooled down. The heat remaining in the heat exchanger is distributed to the space before the blower stops. This amount of heat may cause the space temperature to rise another two to three degrees. This is called overshoot.

The combination of system lag and overshoot can add as much as 5 to 6 degrees to the temperature variation inside the space. See Figure 48-8.

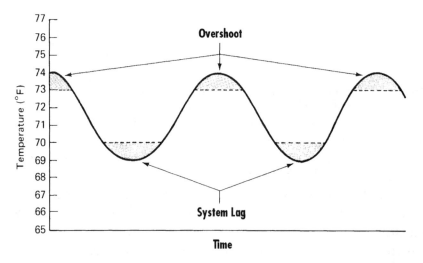

Figure 48-8 Effect of system lag and overshoot on space temperature.

Comfort heating thermostats are equipped with a heat anticipator placed in electrical series with the heating control (gas valve). See Figure 48-9.

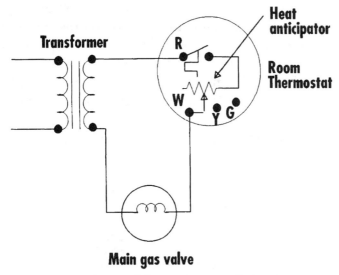

Main gas valve

Figure 48-9 Heating anticipator in control circuit.

This very small resistor is placed near the bimetal so that the heat given off by the anticipator will heat the bimetal, making the thermostat think the space is warmer than it actually is. This causes the thermostat to open the circuit to the gas valve before the space actually reaches the thermostat setting. Heat anticipators are usually adjustable so that they can be adjusted to match the current draw of the temperature control circuit and provide the proper amount of heat to the thermostat.

Operation: In operation, when the thermostat senses a need for heat its contacts will close the electric circuit to the heating control. At the same time electric current passes through the heating anticipator providing a false heat to the bimetal. It will take the furnace a while to heat the heat exchanger before heat is distributed to the space. The fan will come on and distribute the heated air to the conditioned space. The system is now operating normally.

When the thermostat senses that the space is heated sufficiently, its contacts open and interrupt the electricity to the heating control and the heat anticipator. Because the furnace has ran for a shorter length of time it will cool faster and the fan will stop sooner than without the heat anticipator. The furnace heat exchanger is then cooled down and the fan stops. The heat anticipator cools to room temperature. The furnace is now ready for another heating cycle.

Cooling (Off-Cycle) Anticipator: The operation of the cooling anticipator is exactly opposite that of the heating anticipator. The cooling anticipator is connected electrically in parallel with the cooling temperature control circuit. See Figure 48-10.

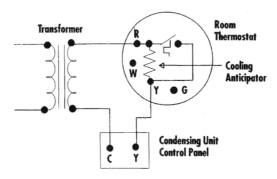

Figure 48-10 Cooling anticipator in control circuit.

It adds heat to the bimetal during the "off" cycle of the cooling system. This will cause the system to come back on sooner than if it were not used. During the "off" cycle electricity flows through both the cooling anticipator and the contactor holding coil. A false heat is added to the thermostat during the off cycle which causes it to think that the space is warmer than it actually is. Because of the series connection, the cooling anticipator reduces the voltage to a point to prevent the contactor from pulling-in. When the thermostat contacts close the electricity takes the path of least resistance and bypasses the anticipator and energizes the contactor holding coil. Cooling anticipators are are sized by the thermostat manufacturer to provide the proper amount of heat and they are generally not adjustable.

System Switch: System switches are placed on thermostats so that the user can select the desired operation. They usually include an "off" position, a "cooling" position, and a "heating" position. See Figure 48-11.

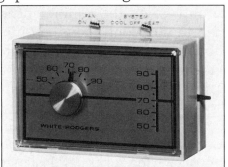

Figure 48-11 System switch and fan switch. (Courtesy of White-Rogers Division, Emerson Electric Company.)

When the switch is in the "off" position none of the equipment except the fan will operate. When the switch is placed in the "cooling" position only the cooling equipment and the fan will operate. When in the "heating" position only the heating equipment and the fan will operate. This allows manual selection for system operation. Some thermostats provide for automatic change-over from heating to cooling and from cooling to heating when the space temperature drops or rises to a predetermined temperature differential. This will provide the most comfort in the space. However, it is also the most expensive operation because with only a very small temperature change some of the equipment will be in operation. **Example;** because these thermostats will automatically change equipment operation on a small temperature change, during the spring and fall of the year the space may need heat in the morning and will be warm enough for cooling in the afternoon. This operation will automatically occur rather than being manually selected. Most of the time the small temperature change remains comfortable enough without requiring that the equipment be switched. This increases the overall cost of conditioning the space.

Fan Switch: The fan switch is placed on the thermostat

so that the operator can select either "automatic" or "continuous" fan operation.

When the switch is placed in the "automatic" position the fan will operate only when the equipment is operating or demands it. When in the "continuous" position the fan will operate continuously and usually in the high speed mode. This is because some people like continuous air circulation and others like intermittent air circulation. Also, installations equipped with electronic air cleaners usually require continuous fan operation to properly clean the air.

Electrical Connections: Almost all thermostats in use today use one of two wiring terminal designations, depending on the brand used. They are as follows:

Purpose	Letter Designation
Transformer Voltage	R or V
Heating	W or H
Cooling	Y or C
Heating Damper (seldom used)	B
Cooling Damper (seldom used)	O

Multi-Stage Thermostats: Staging thermostats are designed to operate different pieces of equipment as the temperature change demands. However, they may also be used to control the compressor unloaders in response to temperature charges inside the conditioned space.

These thermostats are equipped with more than one set of contacts for one or more heating and cooling functions. They are generally termed two-stage cooling, single-stage heating; two-stage cooling, two-stage heating; one-stage cooling, two-stage heating. In some instances there may be more sets of contacts for a given function. Through the different contact arrangements, the thermostat will demand either more or less in response to the temperature demands of the space. Each stage has a definite number of degrees lag before another stage is energized or deenergized. That is there is a temperature differential between each of the stages.

Operation: when the indoor temperature drops during the heating cycle, usually about three degrees below the present temperature of the space, the second stage heating contact on the thermostat will close and complete the circuit to the second stage heating equipment which has the capacity to reach the temperature demanded. When the indoor temperature increases to the cut-out of that stage the equipment is automatically stopped. If the temperature continues to increase, the thermostat will automatically stop all of the equipment just like any other thermostat.

During the cooling cycle, if the temperature rises about three degrees above the present temperature, the second stage contacts in the thermostat will close energizing additional cooling capacity. As the inside temperature falls to the cut-out temperature of the second stage, the thermostat will automatically de-energize the second stage equipment. If the temperature continues to drop, the thermostat will automatically stop operation of the cooling equipment when the thermostat setting is reached. This type of operation is desirable in applications when the load changes a great amount from time to time. It is desirable from both a comfort stand point and an economic stand point. When the system is operating on an unloaded condition it is using less energy than when operating at full capacity. Also, during periods of less cooling demand, the cooling system will operate at the unloaded condition which will keep the refrigeration system operating longer rather than constantly on and off. This continuous refrigeration system operation will help to maintain a lower relative humidity inside the space making it more comfortable.

The wiring connections to these thermostats are labeled the same as single-stage thermostats with the exception of the addition of a 1 and a 2 after each function.

Purpose	Letter Designation
Transformer Voltage	R or V
Heating 1st Stage	W1 or H1
Heating 2nd Stage	W2 or H2
Cooling 1st Stage	Y1 or C1
Cooling 2nd Stage	Y2 or C2
Fan	G or F
Heating Damper (if used)	B
Cooling Damper (if used)	O

See Figure 48-12 for a wiring schematic for a multi-stage thermostat.

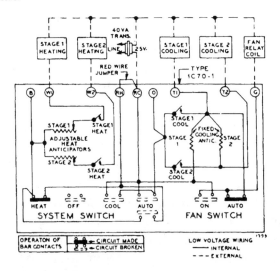

Figure 48-12 Multi-stage thermostat schematic. (Courtesy of White-Rogers Division, Emerson Electric Company.)

Heat pump Thermostat: Most heat pump systems use staging thermostats to energize the auxiliary heat strips when the outdoor temperature drops below the balance point temperature of the heat pump unit. The thermostat first stage operates the heat pump system. When the indoor temperature drops about three degrees below the thermostat setting, the thermostat second stage energizes the auxiliary heat strips to maintain the desired temperature. As the temperature inside the space increases the second stage deenergizes the auxiliary heat strips leaving the heat pump system in operation. When the indoor temperature continues to increase to the off setting of the thermostat deenergizes the heat pump system.

Heat pump thermostats generally are two stage heating and one stage cooling types. The proper type thermostat must be used for satisfactory operation of the heat pump. It is a good idea to use the type designated by the equipment manufacturer. They must be wired according to the thermostat and equipment manufacturers requirements.

Programmable (Set Back) Thermostats:
Programmable thermostats are used to operate the equipment during parts of the day to conserve energy and to bring the space temperature back to normal operating conditions when desired.

These settings may be used for comfort or for energy conservation or both. Set back is a term used to indicate that the thermostat is set back to a temperature that would not normally be comfortable, but will still not require the equipment to bring the space temperature from a maximum temperature difference. They are available in models that simply have a clock that will change the temperature every day at the same time to models that are operated by a quartz-operated time clock and integrated circuits (ICs) which can be programmed for several different settings and for operation on different days of the week. They are available in single-set back and multiple-set back models. The purpose of the clock is to actuate a set of contacts which control the equipment operation at a predetermined time. Some of these thermostats are battery operated and others are connected into the 24 volt control circuit voltage for operation. Some of them operate from the 24 volt circuit and have a battery back-up to keep the clock operating in the event of a power failure. Some multi-stage thermostats may require that transformers be wired in parallel or that an isolation relay be used. Be sure to install the thermostat according to the thermostat manufacturer's specifications.

Operation: Basically, there are two thermostats used to control a function. It is the position of the time clock switch that determines which thermostat is in operation at any particular time.

Lets assume that the system is in the cooling mode with thermostat 1 set at 90 degrees and thermostat 2 set at 75 degrees. When the time clock determines that the space temperature should be at the higher temperature for conserving energy, the clock causes the switch to complete the circuit through thermostat 1. At about one hour before the occupants are to return home the clock causes the switch to change positions and allow thermostat 2 to control the space temperature so that it will be comfortable when they arrive. The clock is usually set to change to the higher thermostat setting at the time, or slightly before, the occupants leave the building. The thermostat operates in the same manner during the heating season.

Thermostat Installation: Install the thermostat and wall plate about 5 ft. above the floor in an area with good air circulation at room temperature.

Do not install the thermostat where it will be affected by

1. Drafts or dead spots behind doors, in corners, or under cabinets.
2. Hot or cold air from ducts.
3. Radiant heat from sun or appliances.
4. Concealed pipes or chimneys.
5. Unheated (uncooled) areas behind the thermostat, such as an outside wall.
6. Do not mount the thermostat where it will be subjected to drafts when a door or window is opened.
7. Do not mount the thermostat on a wall that will be vibrated each time a door is opened or closed.
8. Always mount the thermostat in the space that it is to control.

On a new installation run the wiring cable to a hole at the selected location, and pull about 3 in. of wire through the hole. Color-coded, 18-gauge thermostat wire with at least one wire for each wiring terminal on the thermostat is recommended.

Remove the thermostat from the wallplate. Make certain that the wallplate is level, especially for mercury bulb type thermostats. When leveling is required there are usually leveling marks on the wallplate. The wallplate mounts directly on the wall with screws. Use the wallplate as a template, and with a pencil, mark two mounting screw positions that will fit the application using two of the three mounting holes in the wallplate. Thread the wires through the center opening on the wallplate. Then mount the wallplate. Gently tighten the screws, level the thermostat then securely tighten the screws.

When replacing a thermostat make certain that the correct replacement is used. Otherwise there could be damage to the thermostat or to the equipment. When the old thermostat has provided satisfactory operation until recently, install it in the same manner as the previous one.

Commercial type thermostats should be mounted where they will sense the temperature reflecting the average temperature of the refrigerated space. When being used to control the temperature of chilled or hot water, place the sensing element where the correct temperature is being measured. When installing a thermostat in a commercial refrigeration case or box, the sensing element should be placed where it will sense the average temperature inside the case or cabinet. They must not be installed where they will sense a false temperature such as a warm or cold water pipe or some other source of temperature.

Troubleshooting: Normally thermostats are relatively trouble free controls. However, they do at times require service or maintenance. Thermostats are usually packaged with installation and service information for that particular type control. When service is required there are three areas in which the technician will be concerned: diagnosis, calibration, and routine maintenance.

Trouble diagnosis is generally required when the customer complains that there is an excessive temperature swing inside the space, the system is constantly going on and off, the system operates too long before going off, or that the thermostat is inoperative. When the complaint involves a variation in temperature during the heating season, adjusting the heat anticipator to match the current draw of the heating control circuit will usually solve the problem. During the cooling season when frequent cycling is experienced, check to make certain that the thermostat is not mounted where it will sense drafts or wall vibration. When the thermostat does not function as it should it must be replaced with the correct replacement before normal operation can be realized. If there is a large temperature differential between the on and off of the equipment check for a wide differential between the opening and closing of the contacts. The differential is usually about three degrees. Most thermostats do not have an adjustment for this problem. If it is really a problem the only solution is to replace it.

One of the most common complaints about a thermostat is that it does not maintain the correct temperature inside the space. This generally requires that the thermostat be calibrated, if it is possible to calibrate it. Use an accurate thermometer to check the opening and closing of the contacts with the temperature indicated by the thermometer. Use the thermostat manufacturer's recommendations. If the thermostat operation does not coincide with the thermometer on the thermostat it may be possible to calibrate the thermostat thermometer to read the same as the thermometer used for calibration. If not inform the customer that the thermometer is not accurate.

Thermostat maintenance involves checking the wiring terminals to make certain that they are tight. Cleaning the contacts if they are the open type. Checking thermostat calibration and cleaning the sensing element.

Commercial type thermostats have basically the same problems as do room thermostats. The major difference is to make certain that the remote bulb, if used, is clean and properly secured to the mounting place.

Unit 49: Pressure Controls

A pressure control is a device that will open or close a set of contacts in response to a pressure sensed by the control diaphragm. These are generally grouped into high-pressure and low-pressure controls. Their contacts may open or close on a change in pressure. They act to start or stop a motor in response to the pressure inside the refrigeration system. Pressure controls are available in electro-mechanical and solid-state types.

The electro-mechanical controls are available in both adjustable and nonadjustable types. The solid-state controls are nonadjustable. Some equipment manufacturers use the non-adjustable types to prevent changing their operating pressure and influencing the integrity of the system. The adjustable types may be adjusted at any point within the range indicated on the adjustment scale. Some pressure controls are manufactured with a removable enclosure and some that the enclosure cannot be removed. See Figure 49-1.

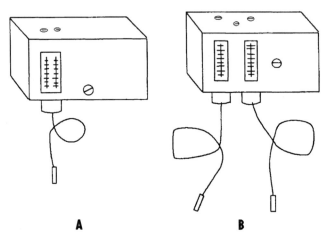

A   **B**

Figure 49-1 Pressure switches-single-and dual function.

HIGH-PRESSURE CONTROLS: These types of controls are generally used as safety devices to protect a motor or some other piece of equipment from damage should the refrigerant pressure inside the system reach dangerous levels. They are connected to the high pressure side of the system. Usually at the compressor discharge where they cannot be rendered inoperative by closing the service valve or some other means. Usually they are wired into the line voltage wiring of the compressor. However, sometimes they may be wired into the low voltage control circuit. See Figure 49-2.

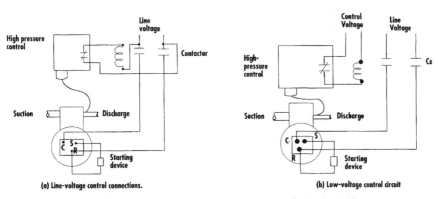

(a) Line-voltage control connections.   (b) Low-voltage control circuit

Figure 49-2 High-pressure control connections.

This will depend on the desires of the equipment manufacturer. They are available in both electro-mechanical and solid-state types. See Figure 49-3.

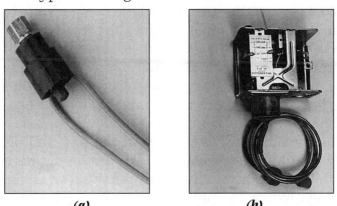

(a)   *(b)*

Figure 49-3 Solid-state (a) and electro-mechanical high-pressure controls (b). [(a) Courtesy of Watsco, Inc.] [(b) Photo by Billy C. Langley.]

High-pressure controls are designed to open the electrical circuit to the compressor when the discharge pressure reaches a predetermined point. This point is chosen by the equipment manufacturer and should not be changed without good reason. Some are equipped with a manual reset and some have automatic reset. High refrigerant pressures can be caused by insufficient condenser cooling, excessive refrigerant charge, air in the refrigerant lines, or some other abnormal condition. High-pressure controls may also be used to start or stop another motor or some type device to help in controlling the cooling of the condenser, or some other desirable operation. The only difference is the range of control operation.

The contacts on a high-pressure control open on a rise in pressure and close on a drop in pressure. They are single-pole, single-throw type switches. When the control used is adjustable it should be adjusted only to provide proper safety to the equipment. The differential between the opening and closing of the contacts may be adjusted on the adjustable types but should not be set close enough together to allow short cycling of the unit.

INSTALLATION: The installation of a high-pressure control is relatively simple. Connect the refrigerant connection where it is not likely to be rendered inoperative by some service operation. Generally between the discharge service valve and the compressor discharge. The electrical wiring is installed where recommended by the equipment manufacturer. Set the control to respond to the desired pressure. To make this adjustment, start the unit and block the cooling medium to cause an increase in the discharge pressure while reading the discharge pressure on a high pressure gauge. When the pressure reaches the desired cut-out point adjust the control to stop the compressor. Remove the cooling medium restriction and place the unit in service.

When replacing a high-pressure control it is best to use an exact replacement. If not possible, use a recommended replacement. Be sure to use one that has the proper scale if it is an adjustable type. If non-adjustable make certain that the cut-out and cut-in points are at the correct pressure.

Troubleshooting: High-pressure controls do not usually cause much trouble. They are usually either good or bad. However, they will sometimes require a slight adjustment to maintain the desired operation. They can be checked by restricting the cooling medium to the condenser while observing the discharge pressure. If the control does not stop the compressor when the desired pressure is reached, adjust the control. If it cannot be adjusted to provide the desired operation, replace it. Never leave a high-pressure control bypassed. To do so will certainly invite trouble because there is little protection from a high pressure condition. It is not recommended that a leak be repaired on high-pressure controls because the capillary tube may be restricted during the soldering process, rendering the control inoperative. When a leak develops in the bellows, the control must be replaced because it is impossible to repair a leaking bellows.

LOW-PRESSURE CONTROL: Low-pressure controls are used as safety devices and in some commercial refrigeration installations as temperature controls. They are equipped with one set of single-pole, single-throw contacts. The contacts make on a pressure rise and open on a drop in pressure. The scale used on low-pressure controls may cover a wide range. They may be used to stop a compressor on a fall in the low-side pressure or to start or stop another piece of equipment in response to the low-side pressure. They are available in either electro-mechanical and solid-state types. See Figure 49-4.

Figure 49-4 Solid-state (a) and electro-mechanical (b) low-pressure controls. [(a) Courtesy of Watsco, Inc.] [(b) Photo by Billy C. Langley.]

(a)   *(b)*

In air conditioning applications, the purpose of the low-pressure control is two fold: it will stop the compressor to prevent frosting of the evaporator when the low-side pressure drops to the cut-out point. Also, they will stop the compressor when the system charge drops to a point that the compressor does not receive the proper cooling from the suction refrigerant. They may be either manual or automatic reset. In commercial refrigeration applications they may be used as temperature controls. They will stop the compressor when the low-side pressure drops to the point to maintain the desired temperature inside the refrigerated case or fixture. Low-pressure controls used in this manner should not be of the manual reset type.

Installation: When installing a low-pressure control make the refrigerant connection to the system where it is not likely to be rendered inoperative by some service operation. The electrical wiring may be either line voltage or low voltage, depending on the purpose and the desires of the equipment manufacturer. See Figure 49-5.

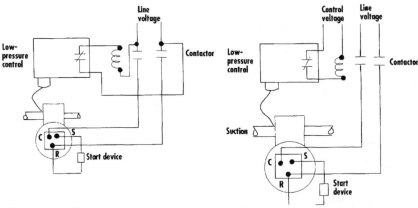

Figure 49-5 Low-pressure control connections.

Make certain that the control stops the compressor when the refrigerant pressure drops to the desired point. This may be done by slowly closing off the low side service valve while observing the low side refrigerant pressure. When the desired cut-out pressure is reached the compressor should stop. If not adjust the control to stop the compressor. It is generally recommended that the cut-in pressure also be set to the equipment manufacturer's recommendations. This can be done by allowing the low-side pressure to rise to the desired cut-in point and adjusting the control if necessary.

Troubleshooting: Low-pressure controls are relatively trouble-free. An occasional adjustment may be needed. When trouble occurs, such as; frosting of the evaporator, high or low temperature inside a refrigerated cabinet just make the simple tests described on making adjustments. If the control does not prove to be operating satisfactorily replace it. Be sure to use an exact replacement or one that is a recommended replacement. Otherwise proper operation may not be possible. Be sure to check the cut-in and cut-out settings of any replacement control. It is not recommended that low-pressure controls found to be leaking refrigerant be repaired because the capillary tube may become restricted during the soldering process. Also, controls with a leaking diaphragm cannot be repaired. When replacing a nonadjustable low-pressure control be sure to check the cut-in and cut-out settings to make certain that they are as desired.

DUAL-PRESSURE CONTROLS: Dual-pressure controls have both the high-and low-pressure controls in one housing. This type of construction provides savings in space, original cost, and electrical wiring. These controls are wired into the electrical system in two ways: **(1)** the control circuit may be interrupted, allowing the contactor to open and stop the motor; or **(2)** main power to the compressor motor may be interrupted, stopping the motor. The method used will depend on the equipment manufacturer's desires. See Figure 49-6.

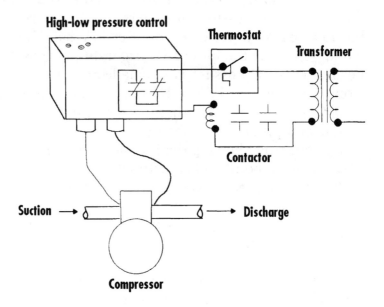

Figure 49-6 Dual-pressure control connections.

Unit 50: Oil Safety Controls

Oil safety controls are safety devices that are designed to protect the compressor against low oil pressure. They are more commonly used on the larger systems. They are available in both electro-mechanical and electronic styles. See Figure 50-1.

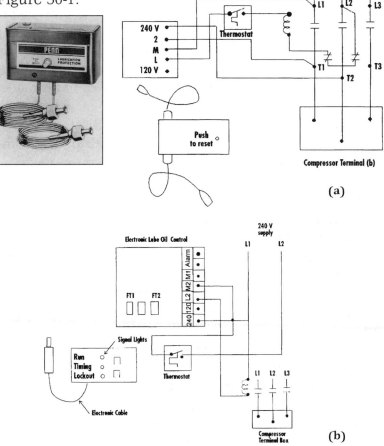

Figure 50-1 Oil safety controls (a) electro-mechanical and electronic controls (b) and schematics.

The larger compressors use an oil pump to deliver the oil to the compressor wearing surfaces under pressure for positive lubrication of all the internal components. There is a built-in time delay switch that allows for oil pressure pickup on starting and avoids nuisance shutdowns on oil pressure drops of short duration during the running and starting periods.

ELECTROMECHANICAL: They operate on a difference in pressure between the oil pump discharge, which will be the sum of the actual oil pressure, and the suction pressure of the system. When the compressor is starting and some times during operation the oil pressure will drop below the control setting. At this point an electric heater in the control is energized. If the oil pressure difference is not again reached and maintained, after a given period of time a switch will open the control circuit to the compressor starter or contactor. As an example, suppose the compressor manufacturer recommended a net oil pressure of 25 lb be maintained, the compressor is operating at a crankcase refrigerant pressure of 45 psig. The oil pump discharge must be 70 psig. Otherwise the oil safety control will stop the compressor. The formula for calculating the net oil pressure is:

Net oil pressure = total oil pressure – suction refrigerant pressure.

The suction pressure is present on both sides of the control diaphragm and therefore cancel each other out. The operation of the control depends on the net oil pressure overcoming the predetermined control spring pressure. The pressure of the control spring must be equal to the minimum oil pressure recommended by the compressor manufacturer.

OPERATION: During operation, the total oil pressure is the combination of the crankcase pressure and the oil pump discharge pressure. The net oil pressure that is available to force the oil through the compressor components is the difference between the total oil pressure and the refrigerant pressure in the crankcase. The control measures this difference which is referred to as the net oil pressure.

When the compressor starts, the time delay switch in the control is energized. Should the net oil pressure not increase to the control cut-in setting within the required time limit, the time delay opens the SPST set of contacts to interrupt the control circuit to the compressor motor starter or contactor to stop the compressor. However, if the net oil pressure rises to the desired cut-in setting within the required period of time after the compressor starts, the time delay switch is automatically deenergized and the compressor continues to operate as designed.

Should the net oil pressure drop below the cut-out setting while the compressor is operating, the time delay switch is energized and, unless the net oil pressure returns to the cut-in pressure within the time-delay period, the control circuit is opened stopping the compressor.

Installation: The time delay function of these controls is compensated for variations in ambient temperature around the unit. Be sure to check to make certain that the control being used falls into the temperature range recommended. Otherwise the control may not operate as it was designed. Any variations in the applied voltage will also affect the timing function of the control. The manufacturer's recommendations should be followed for a proper installation. The control oil pressure line is connected to the connection labeled OIL and the crankcase line to the pressure connection labeled LOW. See Figure 50-2.

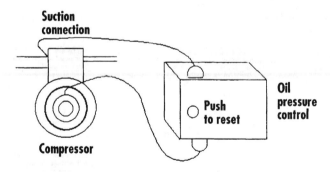

Figure 50-2 Oil safety control connections.

These controls are rated for pilot duty operation only. A typical wiring diagram for a 240 volt control system is shown in Figure 50-3.

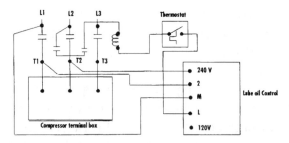

Figure 50-3 240 vac lube oil control wiring diagram.

ELECTRONIC LUBE OIL CONTROL: The electronic lube oil control is used on refrigeration compressors equipped with either a bearing head or an oil pump that will accommodate a single-point differential pressure transducer. They measure the net oil pressure and de-energize the compressor if the oil pressure drops to the control set point. Some are equipped with an LED display to show the condition of the lubrication system. See Figure 50-1. An anti-short-cycling delay is sometimes used. Also, an accumulative or noncumulative timer may be required by the compressor manufacturer.

Operation: The operating cycle of the lube oil control is shown in Figure 50-4.

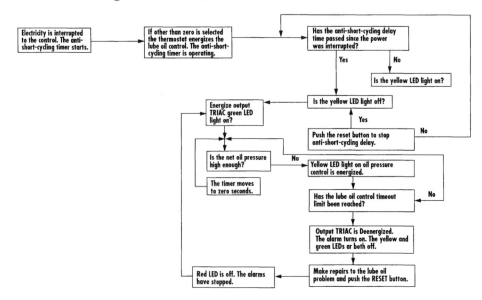

Figure 50-4 Typical electronic lube oil control operation.

Electrical Checkout Procedure: As an example use the following procedure to test for correct operation during the initial installation and maintenance operations of the electronic Lube Oil Control. Be sure to use the manufacturer's recommended procedures for electronic lube oil control being tested.

1. De-energize the supply voltage to the control and the compressor electrical circuit.

Warning: To avoid risk of electrical shock or damage to the equipment, determine if more than one supply voltage exists and disconnect all of them before proceeding.

2. Disconnect the wire leads between the starter and compressor motor "T" terminals as shown on the wiring diagram, see Figures, 50-5 through 50-10 to stop the compressor from running during this part of the test.

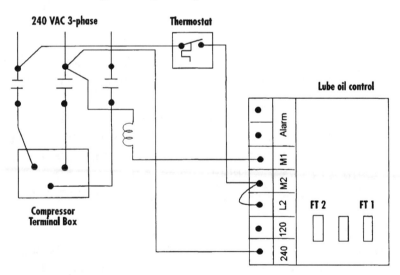

Figure 50-5 Three-wire lube oil control wiring diagram.

Note: On systems using a current sensing relay, remove the relay connections to the control terminals (FT1 and FT2), and connect a jumper between these two terminals. (See Figures 50-9 and 50-10.)

3. Re-energize the supply voltage to the control. Check to make certain that all operating and limit controls are closed to ensure that power is being supplied to the control.

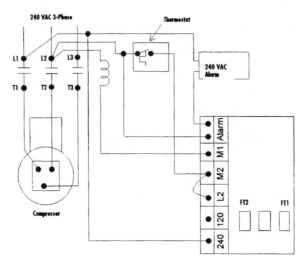

Figure 50-6 *The alarm circuit being powered by the same voltage supply as the motor supply.*

4. The compressor contactor circuit will immediately be energized , the oil pressure differential is low and the timing circuit is energized.

5. In about 45, or 120 seconds, the lube oil control will lockout the starter. If an alarm is installed, the alarm contacts will close and the alarm will sound.

6. Press **RESET.** The starter is now energized.

Note: The control will remain locked out until the RESET button is pressed even if power is removed. The control cannot be reset without power.

7. De-energize the supply voltage. Reconnect the compressor leads to the starter. If a current sensing relay is used, remove the jumper and reconnect the leads to the control.

8. Re-energize the supply voltage. If the operating and limit controls are closed, the compressor will start. The control heating device will turn off when the pressure level reaches the setpoint, generally within seconds of starting the compressor.

Repairs and Replacement: Field repairs must not be made. Sensors, sensor cables, and replacement controls are available as separate items through local wholesalers and the original equipment manufacturer. Always replace the defective component rather than attempting to repair it.

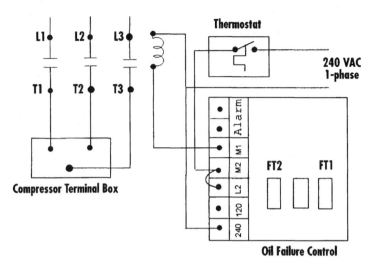

Figure 50-7 Separate voltage used by the control circuit.

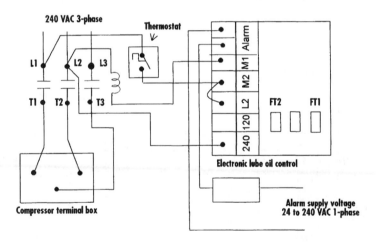

Figure 50-8 Electronic lube oil system with Separate voltage supply.

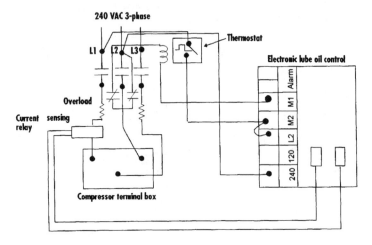

Figure 50-9 Lube oil control operating motor with overloads and a current sensing relay.

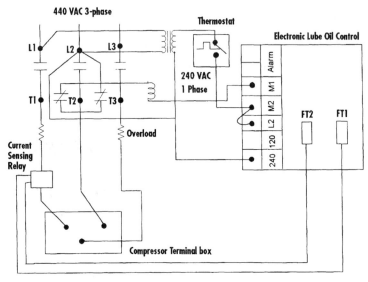

Figure 50-10 A 240 or 480-vac supply using a 240 vac magnetic starter coil and transformer.

Unit 51: Defrost Timers

Most of the refrigeration systems in use require that the evaporator coil be defrosted periodically to keep the efficiency as high as possible and to maintain the desired temperature inside the fixture. These controls are available in both electro-mechanical and electronic types.

Defrost timers are available in both adjustable and nonadjustable types. The adjustable types are used mostly on commercial refrigeration and air conditioning applications. The nonadjustable types are generally used on domestic refrigeration applications. They are also available in electronic models.

The adjustable models are available in either one day or seven day operation. See Figure 51-1.

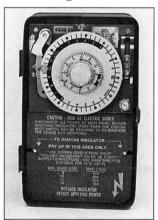

Figure 51-1 One day time clock.
(Courtesy of Paragon Electric Co., Inc.)

The one day clocks have clips, or dogs as they are usually called, that operate through a 24 hour cycle and repeat the same cycle at the same time each day. The seven day time clock can have these dogs set at a given time for a single day or for multiple days. They can also be set to perform a different function on different days or at different times during each day, depending on the requirements of the system being controlled.

Typical wiring diagrams for this type of defrost control are shown in Figures 51-2 and 51-3.

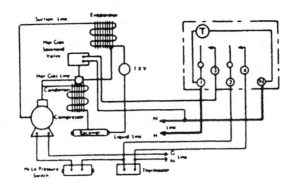

Figure 51-2 Hot gas defrost connections and clock operation. (Courtesy of Paragon Electric Co., Inc.)

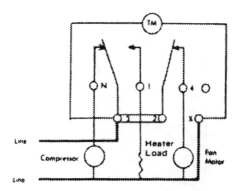

Figure 51-3 Electric heat defrost connections and clock operation. (Courtesy of Paragon Electric Co., Inc.)

The electronic defrost controls are available in multi-circuit design. See Figure 51-4

Figure 51-4 Electronic defrost control. (Courtesy of Paragon Electric Co., Inc.)

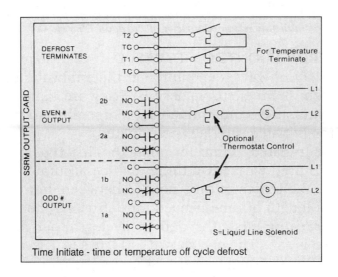

Figure 51-5 Electronic defrost control wiring diagram of hot gas defrost. (Courtesy of Paragon Electric Co., Inc.)

When it is necessary that these controls be field set it is recommended that the manufacturer's instructions be followed. The following schematic wiring diagrams show the systems that they can be used on. See Figures 51-5 and 51-6.

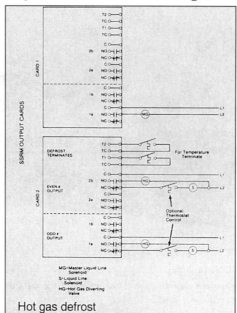

Hot gas defrost

Figure 51-6 Electronic defrost control wiring diagram for a hot gas defrost with fan relay. (Courtesy of Paragon Electric Co., Inc.)

DEFROST CONTROL TYPES: Defrost systems operate in a variety of ways, depending on the application and the equipment manufacturers desires. They are: time-initiated, time-terminated; time-initiated, temperature-terminated; and time-initiated, pressure-terminated.

Time-Initiated, Time-Terminated: This type of defrost control is used where two things can be accurately predicted. One, the length of time needed between defrost cycles. Two, the length of time needed for defrosting the coil.

Operation: First, the defrost cycle tripper on the 24 hour dial is set. (The one shown in Figure 51-1 is set to start the cycle at mid-night, every 24 hours.) If more cycles are needed, additional cycle trippers may be placed on the 24 hour dial. The defrost cycle can be set for as often as every four hours, up to six times a day.

Second, The defrost cycle time set pointer is set on the two hour inner dial. It is set to last for as long as you want the defrost cycle to last. The cycle can be set to last from four to 110 minutes, in two minute increments. The length of defrost is accurate to within ± two minutes.

1. Time initiated, time-terminated compressor shutdown defrosting.

2. Time-initiated, time-terminated electric heat defrosting.

3. Time-initiated, time-terminated hot gas defrosting.

Time-Initiated, Temperature-Terminated: This type of defrost time control starts the defrost cycles according to the times set. However, a rise in temperature terminates the defrost cycle and starts the refrigeration cycles. The temperature control is set at the reading at which all the frost and ice have melted off the coil.

Operation: First, the defrost cycle trippers are set on the 24 hour dial. Next, the fail-safe tripper is set on the inner dial. This control has a defrost period termination fail-safe built into the inner dial. If the equipment is not brought back to a refrigeration cycle by an increase in temperature, the fail-safe will terminate the defrost cycle. A small solenoid (defrost release coil) is built into each control of this type. This solenoid is connected to an external temperature control (cycle limit-switch) which is furnished by the installer. This external temperature control is set to energize the timer solenoid if the equipment is not brought back to a refrigeration cycle.

1. Time-initiated, temperature-terminated compressor shutdown defrosting.

2. Time-initiated, temperature-terminated electric heat defrosting.

3. Time-initiated, temperature-terminated hot gas defrosting.

Time-Initiated, Pressure-Terminated: Like all the others this type of defrosting time control starts defrost cycles according to the time set, except when the coils are free of frost. A pressure bellows on the control which is connected to the suction side of the refrigeration system senses when the coils are frost-free and starts the refrigeration cycle again no matter what amount of time has elapsed. This type of defrost control is used when the compressor is remote from the refrigerated display case.

Operation: First, the cycle trippers are set on the 24-hour dial. Next, the pressure setting is adjusted to the pounds of pressure corresponding to that in a frost free evaporator coil. It may be adjusted from 36 to 110 psi for R12, R22, or R502.

Then the fail-safe tripper is set on the inner dial. This fail-safe device guarantees that the timer will never fail in the defrost side of the cycle. If the pressure bellows does not work, the tripper turns on the refrigeration cycle.

1. Time-initiated, pressure-terminated compressor shut-down defrosting.

2. Time-initiated, pressure-terminated electric heat defrosting.

3. Time-initiated, pressure-terminated hot gas defrosting.

DEFROSTING METHODS: There are several methods used to defrost evaporating coils used in refrigeration applications. They are: compressor shutdown, water spray, hot gas, and electric defrost. A discussion of each of these follows.

Compressor Shutdown: This method is usually used on applications where the space air temperature is 1 degree C (33.8 degrees F) or higher. When the space temperature requirement is satisfied, a thermostat acts to close the liquid line solenoid valve and shut down the compressor. See Figure 51-7.

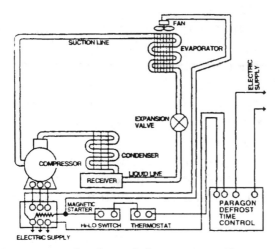

Figure 51-7 Compressor shutdown defrost system. (Courtesy of Paragon Electric Co., Inc.)

The air circulating fans continue to run and the frost on the evaporator coil is melted. In most applications, a defrost time control is used to operate the defrost cycle. These defrost times are generally two hours or more.

Water Spray: This method is usually used for coils operating in room temperatures of -18°C (-0.4°F) and higher. The refrigeration system and air circulating fans are stopped and water is sprayed over the coils. See Figure 51-8.

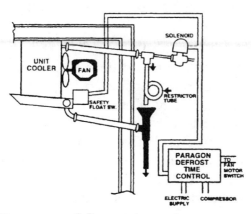

Figure 51-8 Water spray defrost system.
(Courtesy of Paragon Electric Co., Inc.)

It may also be necessary to heat the water supply if the normal water temperatures are below 4.4°C (40°F). The water supply and drain lines should be built to assure rapid drainage of the refrigerated space when the defrost cycle is completed. The defrost time control in this case controls the delay or pump down cycle, the length of the actual defrost cycle in which water is cascaded over the refrigerated coils and the drain time allowing the defrost water to drain away from the refrigerated area.

Hot Gas: This method uses compressed refrigerant vapor from the compressor discharge to apply heat directly to the evaporator and in some systems to the drain pan. See Figure 51-9.

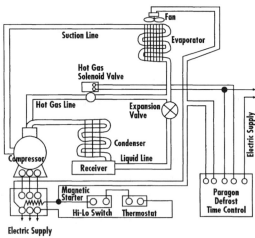

Figure 51-9 Hot gas defrost system. (Courtesy of Paragon Electric Co., Inc.)

Most systems use the latent heat of condensation of the compressed vapor as the heat source, but some use only the sensible heat of the highly superheated vapor. Most hot gas systems introduce the hot gas into the suction connection and bypass the expansion valve through a relief valve into the suction line downstream from a suction solenoid valve which closes during defrost. The defrost time control will, in this case, operate the compressor during the defrost cycle and shut off the circulating fans. At the same time, it will energize the hot gas solenoid valve and allow the hot gas to enter the evaporator coil and warm it, thus removing the build-up of frost.

Electric Defrost: This method uses externally mounted strip heaters that are energized by the defrost time control. These systems, however, require a longer defrost period than hot gas methods, usually 1-1/2 times as long. The heating elements used in supplying the defrost heat may be in direct contact with the evaporator or they may be located between the evaporator fans and the evaporator. In each instance, a temperature limiting device should be used on or near the evaporator to prevent an excessive temperature rise if any control fails to operate. The defrost timer control used in this method will, therefore, turn the compressor off during the defrost cycle and turn on the heaters for the time required to defrost the coil. See Figure 51-10.

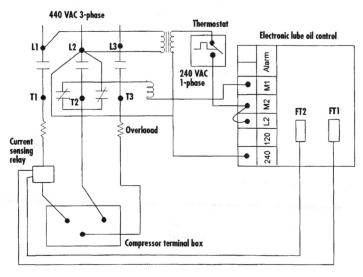

Figure 51-10 Electric defrost system. (Courtesy of Paragon Electric Co., Inc.)

DOMESTIC REFRIGERATOR DEFROST CONTROL: To help maintain the efficiency of the refrigerator a defrost control is also used to direct the removal of frost from the evaporator. See Figure 51-11.

***Figure 51-11 Universal Electronic domestic defrost timer
(Courtesy of Sealed Unit Parts Co., Inc.)***

Domestic refrigerator timers are operated by a single-phase synchronous motor. The same type that is used to power electric clocks. When the timer makes a certain amount of movement a set of contracts are caused to change position, causing the compressor to either operate or stop operating, and to energize or deenergize the electric defrost heating element. These contacts are actuated by a cam that is driven by a gear which is turned by the movement of the timer motor. See Figure 51-12.

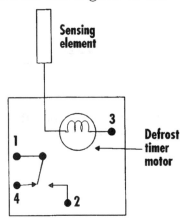

Figure 51-12 Typical domestic defrost timer schematic.

Terminal number 1 is connected to the common terminal of a single-pole double-throw switch with a movable contact attached to it. Terminals 2 and 4 are connected to stationary contacts that direct the power to the proper place at the desired time.

The automatic defrost timer is driven by a self-starting electric motor. The timer motor is wired in series with the control thermostat and runs only when the compressor is energized.

Operation: During the normal operating period an electric circuit is made between contacts 1 and 4. The compressor and all other components needed for the cooling cycle are energized. When defrost is demanded, the movable contact will change position, completing the circuit between terminals 1 and 2. When the timer contacts are in this position the compressor and fan motor are stopped to prevent blowing the warm air inside the refrigerator cabinet.

One lead of the motor is connected to terminal 3 and the other motor lead extends to the outside of the timer case. This configuration allows the timer motor to be connected into the operating sequence of the refrigerator in either one of two ways. **1.** To operate as a continuous run timer. **2.** To operate as an cumulative compressor run timer.

The Continuous Run Timer: As the name suggests a timer connected in this way will run all the time that electricity is applied to it. See Figure 51-13.

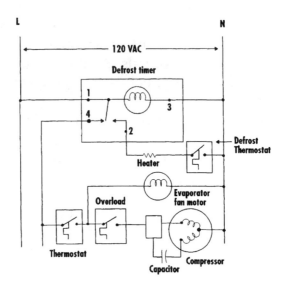

Figure 51-13 Schematic diagram for a continuous run defrost timer.

Notice that the pigtail motor lead is connected to terminal 1 on the timer case which is connected directly to the source of electric power. Terminal 3, the other motor lead is connected to the other side of the power source. When the timer is connected in this manner the motor will operate continuously when electric power is applied to the refrigerator.

During operation, there is an electric circuit through the timer motor and through terminals 1 and 4 so that the timer motor will continue to operate and the compressor and evaporator fan motor will operate when demanded by the thermostat. See Figure 51- 14.

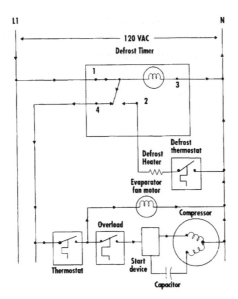

Figure 51-14 Schematic diagram of a continuous run defrost timer when cooling is demanded.

When the timer has advanced enough to demand a defrost period, the timer switches the movable contact from terminal 4 to terminal 2. See Figure 51-15.

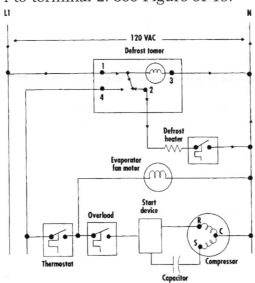

Figure 51-15 Schematic diagram for a continuous run defrost timer when defrost is demanded.

This switching stops the compressor and fan motor but it leaves the timer motor in the circuit for continued operation. At the same time the defrost heater is energized. The heater is warmed to melt the ice and frost from the evaporator. When the defrost timer has advanced enough to terminate the defrost cycle the switch will move the contact from terminal 2 to terminal 4. This change deenergizes the defrost heater and energizes the compressor and fan motor for cooling operation, if demanded by the thermostat.

The Cumulative Compressor Run Timer: When this type of defrost timer is used the timer motor operates only when compressor operation is demanded by the thermostat. The motor is wired in series with the thermostat so that electricity will be provided to it when the thermostat contacts and terminals 1 and 4 on the defrost control are closed. See Figure 51-16.

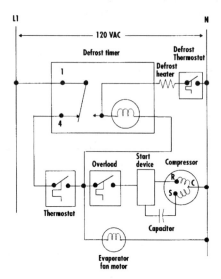

Figure 51-16 Schematic for accumulative compressor run defrost timer.

The timer motor pigtail is now connected to terminal 2 rather than terminal 1.

In cooling operation, the timer contacts 1 and 4 are made, directing the electric current to the refrigerator thermostat. When the thermostat demands cooling, power is directed through the compressor circuit, the evaporator fan circuit, and the defrost timer motor. See Figure 51-17.

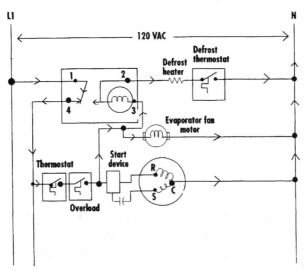

Figure 51-17 Schematic for accumulative compressor run defrost timer during cooling.

Notice that the defrost timer motor is connected in electrical series with the defrost heater. This does not affect the operation of the timer motor because the impedance of the timer motor is much greater than the resistance offered by the defrost heater. Because of this, most of the voltage through the timer and heater is dropped through the timer motor. Because of this reduced current flow through the heater it produces no heat during the cooling operation.

When the defrost timer has initiated a defrost cycle the timer contacts change from 1 to 4 to 1 to 2. See Figure 51-18.

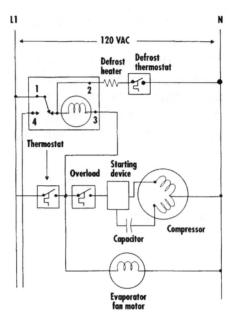

Figure 51-18 Schematic for accumulative compressor run defrost timer during defrost.

The defrost timer is not in series with the defrost heater and full power is supplied to the heater. The heater will now heat and melt any ice and frost from the evaporator. The flow of current is now through the timer motor, through terminal 3, and through the compressor motor circuit. The circuit is through the run winding of the compressor motor. However, the impedance of the timer motor is much greater than the impedance of the run winding of the compressor motor. Again, almost all of the voltage is dropped across the timer motor and causes no action of the compressor motor. When the defrost cycle is completed the defrost timer will change the movable contacts back to their normal cooling position and the system is ready for another cooling cycle.

TESTING A TIMER: It generally takes a lot of time to really test a domestic defrost timer properly. Most timers include a means of manually advancing the timer to start the defrost cycle. Checking from terminal 2 to a neutral point will indicate if the defrost circuit is energized. If it is then wait for the timer to switch the movable contact from terminal 2 to terminal 4. At this point there should be no voltage indicated at terminal 2 but it should be indicated at terminal 4. Also, in this position, if the refrigerator thermostat is demanding cooling the compressor will start. If this checks out, the defrost timer is operating properly. If any part of the test fails replace the timer.

DEFROST TERMINATION CONTROLS: The purpose of defrost termination controls is to stop the defrost period and restart the cooling cycle. There are several types available and the type used will depend on the equipment manufacturers choice. Defrost termination controls are available in temperature, pressure, time, and, time and temperature.

Temperature Defrost Termination Control: Temperature termination controls are used where a certain amount of time will generally remove all the frost and ice from the evaporator coil. They have a sensing element attached at some strategic point, usually the last part of the coil that the ice will melt from, on the coil so that when all the frost and ice is removed the temperature of the coil at that point will start to increase. When the selected temperature is reached the termination control will switch the refrigeration system on, if cooling is needed, and restart the cooling cycle.

This type of termination control is popular on domestic refrigerators and freezers, heat pump systems, and some commercial refrigeration systems.

Operation: The defrost initiation control used with this type of termination control is generally time actuated. When a certain amount of time has lapsed, the time clock will initiate the defrost cycle. When the defrost cycle is initiated, the timer contacts will change position to energize the defrost cycle and stop the cooling cycle. At this time the evaporator fan is stopped to prevent warm air from being blown into the conditioned space. The defrost electrical circuit will keep the defrost cycle on and the cooling cycle off until the defrost termination control switches back to the cooling position. The defrost control contacts will normally reset into the cooling position after a few seconds. When the coil temperature has reached the predetermined temperature, the termination thermostat will automatically switch the electrical circuit back through the defrost control cooling contacts. If cooling is needed at this time the compressor and evaporator fan will start running. The system will operate in the cooling mode until the next defrost cycle is initiated by the defrost timer.

Pressure Defrost Termination Control: Pressure termination controls are generally time initiated control systems. When the desired time is passed the defrost timer initiates the defrost period. There is a connection to the refrigeration system, usually into the low side, to sense the refrigerant pressure. When the prescribed pressure is reached the termination control places the electrical system back into the cooling mode. This type of system is most popular on commercial and industrial refrigeration systems.

Operation: When the time is reached that the defrost timer is set to initiate the defrost cycle, the timer will stop the cooling cycle, and the evaporator fan. The defrost period will begin and the defrost timer contacts will change back to the cooling position, but the defrost electrical system will keep the system in the defrost cycle. When all of the ice and frost has melted from the coil the refrigerant pressure will start to increase. When the pressure reaches a point that indicates the coil is defrosted the pressure control contacts will switch to the cooling position. If cooling is needed at this time the cooling cycle will be started.

Time Defrost Termination Control: Time defrost termination controls operate strictly by the amount of time that has lapsed after the initiation of the defrost cycle. This type of control is usually the most popular on commercial refrigeration systems operating with a case temperature at or just above the freezing temperature of water. In some instances, the time clock that initiates the defrost cycle will also terminate the defrost cycle simply by closing the cooling contacts and opening the defrost contacts.

Operation: When the desired time for a defrost cycle is reached the timer will simply open the electrical circuit to the cooling equipment, the cooling cycle and the evaporator fan will stop. The system will go into the defrost cycle. When the desired amount of time has lapsed since initiation the defrost control will open the defrost contacts and close the cooling contacts. This type of defrost termination control has the fault of sometimes not completely defrosting the evaporator. This will eventually reduce the efficiency and the effectiveness of the refrigeration unit because the frost will continue to build-up on the coil until the air flow is slowed or completely stopped. When this situation occurs, usually a long down period or the manual defrosting of the coil is required. Either method is undesirable because of the time and expense involved.

Time-Temperature Defrost Termination control: These types of systems are most popular on heat pump systems and some commercial refrigeration applications. A defrost timer is used along with a temperature sensor to initiate the defrost cycle. Both controls are required to demand a defrost cycle before one can be initiated. The timer usually closes the defrost contacts about every 30 to 90 minutes. This time is usually adjustable to better maintain the efficiency of the system. The timer will open one set of contacts and close another. If the temperature sensor indicates that a defrost cycle is needed, the system will go into a defrost. If the temperature sensor does not indicate a need for defrost the timer will automatically close the cooling electrical circuit and the system will remain in the cooling mode. They are available with a single bulb temperature sensing element. See Figure 51-19.

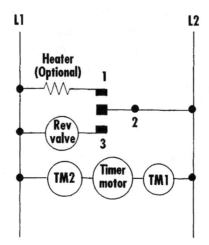

Figure 51-19 Time-temperature defrost control Schematic.

It is the responsibility of the termination thermostat to bring the system out of defrost. The temperature sensors are attached to specific places on the evaporating coil and are insulated. See Figure 51-20.

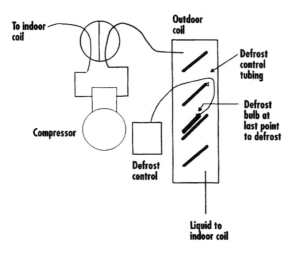

Figure 51-20 Location of temperature bulb for a defrost control.

Operation: In operation, the defrost timer can initiate a defrost cycle only when the evaporating coil temperature is below 30°F and during the time that the defrost timer attempts to initiate a defrost cycle. If the coil is not cold enough the defrost cycle will be skipped for that time period. When the coil temperature is 30°F or lower the defrost initiation will occur. The timer will close the initiation contacts for only a few seconds and then switch back to the cooling position, if the defrost is initiated the system will stay in defrost even though the cooling contacts in the timer have closed. The system will remain in the defrost cycle until the termination temperature has been reached. The defrost termination temperature can usually be set to a temperature that will assure that the coil is completely defrosted. The defrost termination temperature is usually around 60°F. The timer motor usually operates only when the compressor is operating. When the termination temperature of the evaporator coil has been reached and the defrost termination contacts close the cooling system will start. It takes both the timer and thermostat to place the system into a defrost cycle but only the termination thermostat to remove the system from defrost.

Unit 52:
Starters And Contactors

The purpose of a contactor or a starter is to start heavy loads which draw high current. In refrigeration and air conditioning work, the compressor motor represents the largest switching load required of a contactor or starter. The various fan motors, water pumps, and other machinery that may be wired in electrical parallel with the compressor motor also add to the load requirements. There are several methods used to control these various loads but we will discuss only the contactor and starter at this time.

CONTACTOR: A contactor is an electrical device used to complete and interrupt an electrical circuit to a heavy load device. See Figure 52-1.

Figure 52-1 2-pole contactor. (Photo by Billy C. Langley.)

All contactors have basically the same components and features. Each contactor will have a set or sets of contacts that are operated by an electromagnetic coil and armature. There is always a movable and a stationary contact associated with each pole. See Figure 52-2.

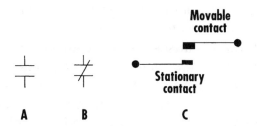

Figure 52-2 (A) Normally open (NO) and (B) normally closed (NC), and movable and stationary contacts.

Some poles will have one stationary and two movable contacts, depending on the desires of the manufacturer. These contact poles may be either open or closed. See Figure 52-3.

Figure 52-3 Contact symbols for energized and deenergized positions.

The **NO** and **NC** designations mean that the contacts are in that position when the contactor coil is deenergized. When a set of contacts are **NO** they will close when the contactor coil is energized. When a set of contacts are **NC** they will open when the contactor coil is energized.

Contactor coils may be energized by the same voltage that they are controlling, such as in commercial refrigeration systems. However, the control voltage for the coil is usually 24 vac for air conditioning. Usually the control voltage is separated from the voltage it is controlling for safety reasons. The current rating of the contactor must be adhered to or the contacts will quickly burn and pit. The cost of replacement will be excessive. The symbols that are used for the coil are shown in Figure 52-4.

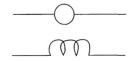

Figure 52-4 Starter or Contactor coil symbols.

Operation: The operating control will complete the control circuit to the contactor coil. The electrical current flowing through the coil causes an electromagnetic field within the contactor coil. This electromagnetic field attracts the armature and pulls it into the center of the coil. The movable contacts are placed on the armature and mate with the stationary contacts of the pole set. The heavy electrical current then flows through these contacts to the compressor motor. The motor starts operating. When the operating control is satisfied it interrupts the control circuit to the contactor coil. The electromagnetic field inside the coil stops and the armature is taken out of the center of the coil by a spring. The movable contacts are moved with it, opening the circuit to the motor. The motor stops and the system is ready for another on cycle when the operating control demands it.

STARTER: The purpose of the starter is basically the same as the contactor with the exception that it will usually include some type of auxiliary devices, such as; overload relays, auxiliary switching contacts, and other devices as desired for proper system operation. They are usually used for larger motors.

Operation: Starter operation is the same as the contactor. The armature is attached to the movable contacts which are caused to move by the electromagnetic field inside the coil. They usually have two or more pole sets to accomplish the desired operation.

They are shown in a schematic diagram like the contactor except that the auxiliary devices are also shown.

Unit 53: Humidistats

In certain air conditioning applications, the control of humidity is of vital importance. Some industrial processes also require that humidity be kept under control for the process involved. The purpose of the humidistat is to control this humidity within the desired limits. See Figure 53-1.

Figure 53-1 Humidistat. (Courtesy of White-Rogers Division, Emerson Electric Company.)

The sensing element of a humidistat must be able to detect very small changes of humidity in the air. This moisture sensitive element will move the mechanical components of the humidistat in response to the amount of humidity sensed to either open or close a set of contacts in the control circuit. The sensing element usually consists of human hair or nylon. These materials will either expand or contract with a change in humidity. It is this expansion and contraction that causes the contacts to open or close. There is an adjustment dial located on the front of the humidistat so that the humidity range can be easily changed to suite the application.

When humidification is used with a central heating or cooling system, the humidistat is wired into the system so that it will operate only when the indoor fan is operating, usually during the heating season. This may be accomplished with either a sail switch located in the air stream, or wired so that it will operate anytime the fan is operating if humidity is needed in the structure. See Figure 53-2.

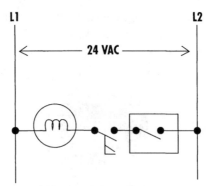

Figure 53-2 *Typical humidistat wiring diagram.*

Should the humidifier operate without the fan operating the humidity would not be properly distributed throughout the building which could cause problems with the equipment rusting and the building would not be properly humidified.

OPERATION: When the humidity inside the conditioned space falls below the on setting of the humidistat, the sensing element senses this condition and causes the humidistat contacts to close the control circuit to the humidifier water solenoid. If the fan is not running, the solenoid will not be energized because of the electrical interlock between the fan and the humidifier. When the fan starts, the interlock circuit is completed and the humidifier solenoid is energized allowing water to flow into the humidifier pads. If a sail switch is used as the interlocking device the humidifier solenoid circuit will not be completed until the air flow reaches the required volume. If the solenoid is interlocked with the fan control it will start when the fan is energized. When the humidity inside the space increases to the off setting of the humidistat, it's contacts will open and deenergize the water solenoid stopping the flow of water to the humidifier.

Unit 54: The Transformer

In cooling and heating systems, transformers are used to reduce the line voltage to low voltage, or 24 V, for the temperature control circuit. They are also used to reduce the voltage from the distribution system to the voltage used in residential and commercial buildings. Transformers may also be used to increase the voltage from one system to another.

A transformer may be defined as a device used to transfer electrical energy from one circuit to another by electromagnetic induction. A transformer consists of two or more coils of wire wound around a common laminated iron core. Thus, the coupling of the two coils can almost reach unity. Unity is considered to be one and is considered to be a perfect coupling between two coils. Transformers have no moving parts and, therefore, require little maintenance. They are simple, rugged, and very efficient devices.

CONSTRUCTION: The construction of a transformer is very simple. The input winding is called the primary winding. See Figure 54-1.

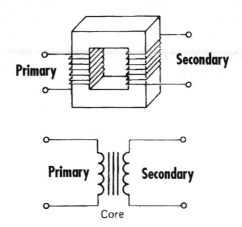

Figure 54-1 Transformer construction and symbol.

The primary winding is connected to the power source. The output winding is called the secondary winding. The output load is attached to the secondary winding. The electrical energy in the secondary winding is the result of mutual induction between the primary and the secondary winding. The magnetic field in the primary winding must be varying and cut across the wires in the secondary winding to induce a second voltage. Because of this requirement, a transformer must be connected to an ac or pulsating dc electrical supply.

TRANSFORMER TYPES: There are several types of transformers in use today. Probably the two most popular types are the iron-core and the air-core types. They are named from the type of material used between the primary and secondary windings. In the air-core transformer, the flux lines follow an air path between the two windings. The coupling in these transformers is not as good as that in the iron-core types. The iron-core type is used almost extensively in air conditioning and refrigeration control systems.

These transformers have a coefficient of coupling of around 0.98%.

TRANSFORMER SIZES: The transformers used in air conditioning and refrigeration systems provide 24 V at the secondary winding connections and vary in power (volt-amperes) output. The selection of a transformer is very important. It may be larger than needed, in volt-ampere rating, but it must never be smaller than needed or it will burn out due to an overload condition. Notice that the output power is rated in volt-amperes, commonly referred to as VA, which is a product of the secondary voltage and the amp capacity. This VA capacity is stamped on the transformer. See Figure 54-2.

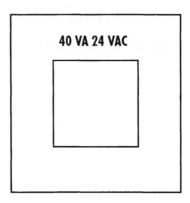

Figure 54-2 Transformer VA rating.

Example: We have a transformer that is stamped 20 VA capacity, and its secondary is 24 V. Divide the capacity, 20 VA, by the secondary voltage, 24 V. This gives us a capacity of 8/10 of an ampere.

Most of the components used in air conditioning and refrigeration control systems are rated in amperes of current draw. Thus, if we were to install a gas valve in a furnace with a current draw listed in amperes, multiply the amperes, in our example 0.45 A, times the voltage, 24, to find the VA capacity needed for transformer selection. Thus, 0.45 x 24 = 10.8 VA capacity. This indicates that a transformer rated for more than 11 VA would do the job. However, the next size usually available is 20 VA, which would handle the load with a little power to spare. The control transformers used in air conditioning and heating systems generally range from 20 to 77 VA and may have either 120 VAC or 240 VAC primary connections. Some transformers are equipped with dual voltage primary windings.

When replacement is required, be sure to replace it with one that has sufficient VA capacity to handle all the loads that may be operating at the same time. It should be remembered that control transformers with a capacity of more than 40 VA be provided with some type of primary overload device, such as a fuse, to prevent damage to the control circuit components should a short occur in the control voltage system.

Unit 55: Variable-Speed Motor Controller

With the attention given to comfort and economy, many manufacturers have started using variable controllers for their motors. These controllers are used to control the speed of condenser fan motors to maintain the proper head pressure when the compressor is running. The controllers are used on the indoor fans to maintain the desired air circulation through the conditioned space to provide the desired comfort. The speed of the motor is changed automatically to maintain the desired comfort and efficiency. They are available in both electric and electronic models.

The two motors used for speed control are the shaded pole and the permanent-split capacitor (PSC) types. These types are popular because there are no special requirements that a start winding be energized by a starting relay or a centrifugal switch as is required for the other motors. Motors that are to be used with variable-speed controllers have windings that are equipped with a high impedance stator windings. The high impedance winding limits the current flow through the motor when it is operating at reduced speeds. Speed control may be accomplished by controlling the amount of voltage to the motor, by placing an impedance in series with the motor winding, or by an electronic speed control.

VARIABLE VOLTAGE CONTROL: There are several methods use to vary the amount of voltage to the motor. They are; by using an autotransformer, triac voltage control, and series impedance in the motor circuit.

Autotransformer: This type of transformer has several taps that provide the reduced voltage to the motor winding. See Figure 55-1.

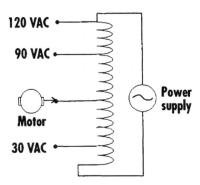

Figure 55-1 Motor voltage controlled by an autotransformer.

This controller has four speeds and an off position. The line voltage is applied across the complete transformer winding. The speed is selected by changing the position of a rotary switch that opens and closes contacts in the circuit. The speed is changed by either placing more resistance in series with the motor winding or removing resistance from in series with the motor winding. More resistance reduces the voltage and causes the motor to run slower. Likewise, less resistance increases the voltage and allows the motor to run faster. When the rotary switch is placed in the 30 volt position the entire transformer winding is in series with the motor winding allowing less voltage to reach the winding which causes the motor to run at the slowest speed for this application. As the rotary switch is moved up through the transformer taps the motor is caused to run at faster speeds because of the voltage increase. The autotransformer provides a separate speed for each tap.

Triac Voltage Speed Control: Triac voltage controls provide a smoother operating motor because the voltage is supplied from zero to full line voltage without incremental settings. See Figure 55-2.

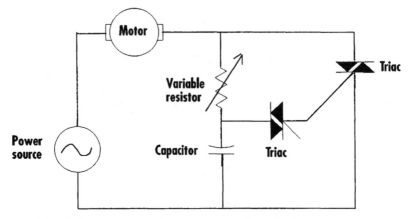

Figure 55-2 Schematic of Triac motor speed control.

The variable resistor shown in this figure is used to control the amount of phase shift for the triac. The triac controls the amount of voltage supplied to the motor by turning on and off at different times during the ac voltage cycle. Use only triac controllers that are designed for motor speed control to prevent damage to either or both the motor and controller.

Series Impedance Motor Speed Control: Motor speed can be controlled by placing an impedance in series with the motor winding. See Figure 55-3.

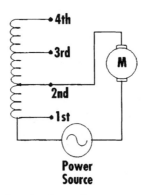

Figure 55-3 Schematic of series impedance motor speed control.

This speed control uses a tapped inductor wired in series with the motor winding. A rotary switch is used to select the motor operating speed. As the switch is moved from one speed to another more or less of the inductor winding is placed in series with the motor winding. When the motor is first started, full voltage is supplied to the motor winding. The rotary switch is then placed in the desired speed position.

Electronic Speed Control: These types of speed controls are usually used on condensing unit fan motors. They may be either temperature or pressure operated. See Figure 55-4.

Figure 55-4 Solid-state discharge pressure control.
(Courtesy of International Controls and Measurements.)

Their purpose is to maintain as constant a head pressure as possible so that the efficiency of the equipment can be maintained in low ambient temperatures. The pressure connection is made at the compressor discharge and the value of the pressure is transmitted to the control through the proper wiring. See Figure 55-5.

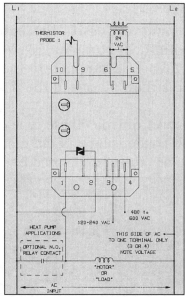

Figure 55-5 *Typical solid-state discharge pressure control ladder diagram (Courtesy of International Controls and Measurements.)*

Some also have a direct pressure connection between the refrigeration system and the controller. Other types use a connection on the liquid line to sense the temperature of the line and control the motor speed in response to this temperature. They provide a continuous speed change to maintain the desired discharge pressure. See Figure 55-6.

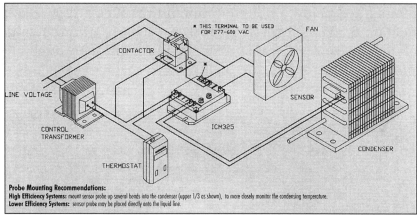

Figure 55-6 *Typical solid-state discharge pressure control installation schematic. (Courtesy of International Controls and Measurements.)*

These are linear type controls in that they do not have steps to control the motor speed. This prevents the pressure from rising and then dropping within the range of various speed taps.

Unit 56: Flow Switches

Flow switches, also known as sail switches, are used in air conditioning systems to detect air flow. Flow switches are sometimes used to detect the flow of a fluid in a pipe. The switch is equipped with a large flat surface (sail) that is placed in the moving stream. When the flow reaches the predetermined rate the sail will move to position a set of contacts inside the switch. See Figure 56-1.

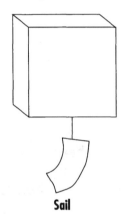

Sail

Figure 56-1 Flow (Sail) switch.

The contacts are enclosed in a snap-action microswitch and are intended for pilot duty use only.

PURPOSE: The purpose of the flow switch is to detect a predetermined flow of air through the space being monitored. It will signal the remaining parts of the control circuit if sufficient air is flowing to maintain adequate operation of the equipment. The flow switch is the only foolproof method of determining if air flow is present. Some manufacturers use a sail switch in both the indoor fan circuit and the outdoor (condenser) fan circuit. This is to make certain that the compressor will not operate until sufficient air flow is detected at both locations. See Figure 56-2.

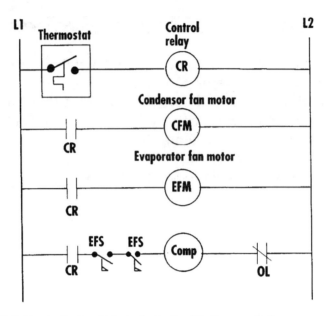

Figure 56-2 Control circuit interlocked with flow switches.

Operation: When the thermostat demands cooling the control relay **(CR)** coil is energized. All of the contacts in the CR actuate, usually close, when the coil is energized. The first set of CR contacts close to energize the condenser fan motor **(CFM)** relay coil to close its contacts and start the condenser fan motor. When the second set of CR contacts close the evaporator fan motor **(EFM)** relay coil is energized and closes its contacts to start the evaporator fan motor. The third set of contacts in the CR cannot energize the compressor contactor coil until sufficient air flow is detected by both of the flow switches because their contacts are in series with the compressor contactor coil. When the air flow is sensed the compressor will start. Should either of the flow switches detect a drop in the air flow through the component it is monitoring, the switch contacts will open and stop the compressor. This is to prevent damage to the compressor from either a low or a high refrigerant pressure.

Unit 57: Fan And Limit Controls

Fan and limit controls have two separate functions within the forced air heating system. They may be in single independent cases or they may both be placed in one case. This will depend on the desires of the equipment manufacturer. When they are both in the same case they are called combination fan and limit controls. They are designed to provide safe and economic heating.

DEFINITIONS: The field accepted definitions of these controls are as follows:

Fan Control: By definition the fan control is a control designed to start and stop the indoor fan in response to the temperature within the heating unit. The contacts are NO SPST and are actuated by the temperature of the air inside the furnace. They are designed to be placed in an exact spot within the furnace heat exchanger so that the air flowing through the heat exchanger passages can be properly monitored by the sensing element of the control. An exact replacement should be used to maintain the accuracy of the furnace design.

Limit Control: By definition the limit control is a safety device designed to interrupt the control circuit to the gas control should the temperature within the furnace heat exchanger increase to an unsafe temperature. The sensing element is placed in the hot spot inside the furnace heat exchanger to control the temperature at this point. An exact replacement should be used to maintain the accuracy of the furnace design. The contacts are NC. Depending on the control design and function, the contacts may be SPST or DPDT that are activated by temperatures within the furnace heat exchanger.

FAN CONTROL: There are several different types of fan controls in use. Regardless of their design their only function is to control the operation of the fan motor in a forced-air heating system. When the thermostat demands heating the gas control is energized. When the inside of the heat exchanger has reached the desired temperature the fan starts running to distribute the warm air to the conditioned space. When the thermostat is satisfied, the gas control is deenergized to stop the heating process. As the heat exchanger cools down the fan control stops the fan motor. This operation is desired to prevent blowing cool air into the space. These controls usually include a temperature dial that can be set at different temperatures to satisfy the requirements of the users. Types: There are basically two types of fan controls used, the temperature actuated and the electrically operated type.

Temperature-Actuated Fan Control: These controls use a bimetal strip, or disc. See Figure 57-1.

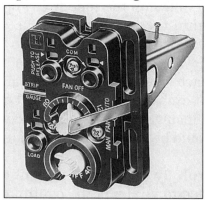

Figure 57-1 Bimetal operated fan controls. Courtesy of White-Rogers division of Emerson Electric Company.

A bimetal is made from two different types of metal that have different expansion and contraction rates. They are welded together so that they respond as one piece. See Figure 57-2.

Figure 57-2 Bimetal.

The bimetal is connected to a set of NO snap acting contacts. The bimetal sensing element is inserted into the circulating air passage of the furnace heat exchanger to sense the heat at the hottest point. If the fan control is replaced be sure to use an exact replacement or the control may not function as desired. The replacement control should have the same element length as the one being replaced so that the temperature will be sensed as designed. They are usually adjustable with a temperature scale located for approximate temperature settings. Some of the bimetal disc types are not adjustable. Fan controls are usually NO SPST type switches that are designed for use on line voltage.

The fan control is wired into the circuit as shown in Figure 57-3.

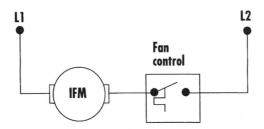

Figure 57-3 Wiring schematic of a heat-actuated fan control.

Operation: When the thermostat demands heat, the heating control is energized to start the heating process. When the temperature inside the heat exchanger reaches approximately 130°F, the fan control bimetal causes the contacts to close and start the fan motor. When the thermostat is satisfied, the heating control is deenergized which allows the heat exchanger to cool down. When the temperature inside the heat exchanger reaches approximately 100°F the bimetal will open the control contacts to deenergize the fan motor. The heating unit is now ready for another cycle when the thermostat demands.

Should the temperature inside the heat exchanger continue to rise to an unsafe level, the limit control will open the control circuit to the heating control and stop the heating process. The fan will continue to operate until the heat exchanger has cooled enough to allow the limit control to close the control circuit and start the heating process again. This situation could occur when the air filter is clogged or when the air flow has been reduced to prevent enough air to be circulated to keep the furnace operating. The cause for this problem must be found and corrected.

Electrically Operated Fan Control: This type of fan control is sometimes referred to as a fan timer control. See Figure 57-4.

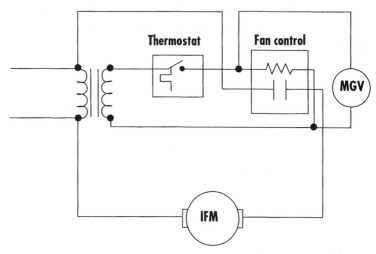

Figure 57-4 Schematic wiring diagram for an electrically operated fan control.

This control operates from heat generated inside the control by a small resistance heater wired into the control circuit in parallel to the heating control. The bimetal inside the control is connected to a set of NO SPST snap acting contacts. These types of fan controls cause the fan to operate on a timed cycle rather than sensing heat inside the heat exchanger. This type of control is very suitable for use on counter-flow, horizontal-flow furnaces because of the peculiar heat build-up in these furnaces, and for electric furnaces because of the rapid heat-up of the heating elements. The heat-actuated fan control will not usually respond fast enough to prevent the high-limit control from stopping the heating process on electric furnaces.

There are four connections on this type of fan control. Two are for connection to the 24 volt control circuit and two are for connecting the fan motor to the line voltage circuit. Refer to Figure 57-5.

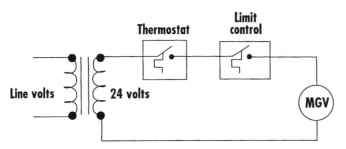

Figure 57-5: Schematic of limit control in 24-vac circuit.

The fan control contacts are usually caused to close at about one minute or less after the thermostat demands heating to start the fan motor. This allows the heat exchanger or heating elements to warm sufficiently to prevent blowing cool air into the conditioned space. When the thermostat is satisfied the fan control contacts will open in about 2 minutes to stop fan operation.

The time delay will be determined by the desires of the equipment manufacturer. They are available in a variety of on and off times. One of the problems with this type of fan control is that the fan motor will operate regardless of whether or not heating is in process, allowing cool air to be blown into the space.

Operation: When the thermostat demands heat, both the heating control and the heater inside the fan control are energized at the same time. The heating process begins almost immediately. After the required time delay of the fan control the fan motor is energized to blow warm air into the space. When the thermostat is satisfied, the control circuit is opened which deenergizes both the heating control and the heater inside the fan control. The heating process stops almost immediately but it takes about two minutes for the bimetal inside the fan control to cool enough to cause the contacts to open and stop the fan motor. The furnace is now ready for another heating cycle when the thermostat demands it.

LIMIT CONTROLS: Limit controls are safety switches and are not used as operating controls. That is they operate only when an unsafe high temperature condition is present inside the furnace. They are bimetal operated temperature sensing controls that are operated by a bimetal. They may be equipped with either NC SPST or SPDT snap acting switches. The SPST contacts open on a rise in temperature inside the furnace heat exchanger. The SPDT contacts will open a set of NC contacts and close a set of NO contacts when the temperature inside the heat exchanger reaches an unsafe value. When the SPST type switch is used, only the control circuit to the heating control is opened when the control actuates. When the SPDT type is used the NC contacts open and interrupt the control circuit to the heating control. The NO contacts close to complete a circuit to the fan motor when the switch actuates. The SPST type contacts may be rated for either pilot duty or for line voltage applications. However, when the SPDT type is used the contacts must be rated for line duty to accommodate the fan motor voltage and amperage.

Types: Limit controls are temperature actuated. They are available with either a bimetal blade sensing element or a bimetal disc sensing element. Some are manually reset and others are automatically reset. They are also available in adjustable and nonadjustable types. The adjustable types have a temperature scale to indicate the approximate temperature setting. The limit control should never be set at a temperature higher than 200°F. Or what the equipment manufacturer recommends. They normally have a fixed temperature differential of about 25°F. which is not usually adjustable.

Operation: When the thermostat demands heat, the heating process is started. When the temperature inside the heat exchanger reaches the fan on setting the NO contacts will close to start the fan operating. When the thermostat is satisfied the heating control is deenergized and the heating process is stopped. When the furnace cools to the fan off setting the fan control contacts open to stop fan operation. Should the temperature inside the heat exchanger continue to rise for some reason, the limit control NC contacts will open to interrupt the circuit to the heating control. The heating process will stop and the fan will continue to operate to cool the heating unit down to the limit control on setting. At this time the heating control is energized and the heating process is started. The heating unit will operate this way until the problem is corrected, unless the limit control is a manual reset type, then the fan motor will run until the heating unit has cooled enough for the fan control to stop the fan motor. The unit will remain off until the limit control is reset. However, the problem will remain and the cycle will repeat itself until the problem is found and corrected.

Solenoid valves are also used in refrigeration systems to either stop or allow the flow of refrigerant through the system on demand from some other operating control. They are also used to control the flow of many fluids such as water, refrigerant, and fuel oil. They may be defined as a component containing two devices: a solenoid coil and a valve. See Figure 57-5.

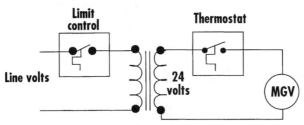

Figure 57-5 Schematic of limit control in line voltage circuit.

The limit control may also be wired so that it will interrupt the line voltage to the primary side of the transformer to stop the heating process. See Figure 57-6.

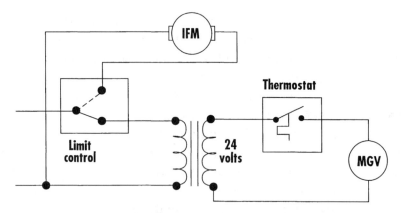

Figure 57-6 Schematic of limit control with SPDT contacts.

In either place the limit control will interrupt the electric power to the heating control. The heating process is stopped until the heat exchanger has cooled enough to allow the limit control contacts to reset, or to be reset, and start the heating process again.

The SPDT type limit control will operate a little different. When the heat inside the furnace has increased to the off setting temperature, the NC contacts will open and the NO contacts will close. This operation stops the heating process and also starts fan operation to remove the excess heat from the furnace faster. See Figure 57-7.

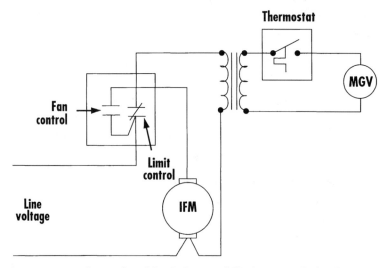

Figure 57-7 Schematic of both fan and limit controls in line voltage.

COMBINATION FAN-LIMIT CONTROLS: Controls of this type provide both functions of fan and limit control. They are bimetal operated and respond to the temperature inside the heating unit. They have snap acting switches and are designed for use on line voltage, 24 volts, or both. The fan control contacts are NO and the limit control contacts are NC. Each control provides the type of operation as if it were a single component. There is an adjustable scale that will allow setting the fan and limit temperatures at the desired values. However, the limit control off temperature cannot be set below the fan control on setting. To do so would effectively remove the fan control from the circuit and the fan would not operate. See Figure 57-7.

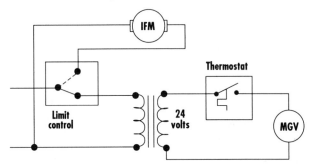

Fig. 57-7: Schematic of limit control with SPDT contacts.

The two switches in these controls must never be wired in series. The limit control may be wired into the low voltage control circuit or into the line voltage to the transformer primary. This will depend on the desires of the equipment manufacturer. The fan control is always wired into the line voltage to the fan motor. See Figure 57-8.

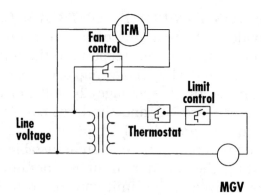

MGV

Figure 57-8 Schematic of fan control in line voltage and limit control in low voltage circuit

Unit 58: Solenoid Valves

A solenoid is a coil of wire that when carrying electricity through it a magnetic field surrounds the coil which in turn causes a valve plunger to move in the desired direction. See Figure 58-1.

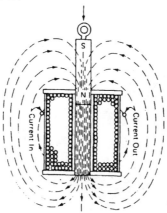

Figure 58-1 Solenoid coil.

Operation: When electric current is applied to the coil a magnetic field is built up around the coil which causes the valve plunger to be pulled into the center of the coil. When the electric current is interrupted, the magnetic field collapses and the weight of the solenoid valve pulls the plunger out of the coil. **Caution:** Electricity must never be applied to a solenoid coil that does not have a plunger inside it. To do so will probably cause the coil to burn-out.

Solenoid valves must always be installed with the arrow pointing in the direction of fluid flow. Otherwise, the valve will probably not open and close properly. They are available in almost any voltage desired for the application. They are also available in both normally open and normally closed types. Be sure to choose the correct one for the application. notice however, the the limit control never interrupts the electrical circuit to the fan motor. If the fan motor were interrupted the furnace would cool down very slowly increasing the possibility of damage to the heating unit or to

the building.
See Figure 58-2.

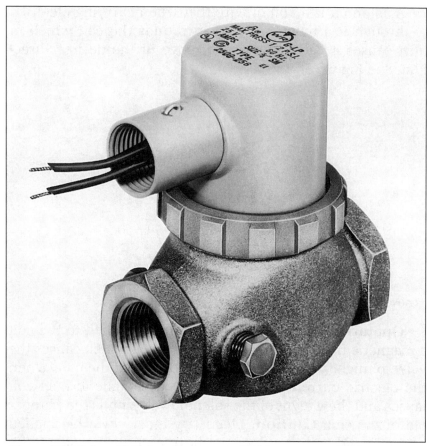

***Figure 58-2 Solenoid Valve. (Courtesy of White-Rogers Division
Emerson Electric Company Inc.)***

Unit 59: The Short-Cycle Timer

Short cycling is a condition that occurs when the compressor is continuously starting and stoping at short intervals. This is a dangerous condition that should not be allowed to continue because it will eventually damage the compressor motor, or any other motor that is affected by the cycling condition. When a compressor has stopped it should be allowed to remain stopped until the pressures between the high- and the low-side of the system have equalized or at least almost equalized. This will reduce the starting load on the motor and help to prevent liquid refrigerant and oil pumping by the compressor.

Time delay controls are used to prevent short-cycling of air conditioning and refrigeration systems to protect the equipment. These controls are available in both thermal and time-delay models. The time delay models are the most popular because they operate much faster than the thermal type and therefore provide greater safety for the equipment. Time delay relays are available in time-delay-on-break and time-delay-on-make.

TIME-DELAY-ON-BREAK: This type of time delay relay prevents the compressor motor from cycling off due to a momentary interruption of the control circuit power. This will give the control system time to reset itself before the system is turned off. See figure 59-1.

Figure 59-1 Time-delay-on-break relay. (Courtesy of International Controls and Measurements.)

Some of the relays are factory set and cannot be changed in the field. Some of them may be adjusted in the field to provide more or less time delay before stopping the system. They are usually adjustable for 3, 4, and 5 minutes. The factory set models are usually set for 5 minutes, or what ever the equipment manufacturer designates.

TIME-DELAY-ON MAKE: Time-delay-on-make relays are used to prevent the system from restarting until it has been off for a predetermined amount of time. Anytime the compressor motor is stopped it will remain off until that time period has lapsed. See Figure 59-2.

Figure 59-2 Time-delay-on-make relay. (Courtesy of International Controls and Measurements.)

When the compressor is stopped inadvertently because of a temporary power outage or for intermittent opening of the control circuit the compressor will remain off until the required amount of time has lapsed, usually 5 minutes. At that time the compressor contactor is energized through the time delay relay and the system is again started for normal operation. This will occur for each power or control circuit interruption. Also the compressor will remain off for the required time period during a normal cycle condition. This is to reduce the starting load on the compressor motor by allowing the high- and low-side pressures to equalize or almost equalize. The time delay period starts after the power is restored to the unit.

Unit 60: Reversing (Four-Way) Valves

These valves are designed to provide four paths for flow-two at a time. The two separate flows can be directed in two different directions because of the valve design. Reversing valves are used on heat pump systems, refrigeration systems to provide a reversed flow of the refrigerant through the system. They use a solenoid valve to actuate the piston inside the valve to direct the refrigerant flow in the right direction at the desired time. See Figure 60-1.

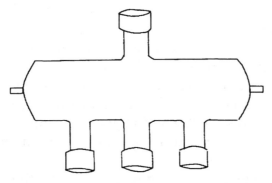

Figure 60-1 Reversing (four-way) valve.

Each valve is designed for various refrigerant flow requirements. These valves are hermetically sealed and are operated by the pressure passing through the solenoid valve.

OPERATION: Basically the reversing valve is made of two separate valves. When the pilot solenoid valve is closed the refrigerant flow is directed in one direction and when it is open the refrigerant is directed in another direction. On refrigeration systems the pilot valve solenoid is generally deenergized during the cooling mode and energized for the defrosting mode. When the pilot solenoid coil is energized on a heat pump system depends on the desires of the equipment manufacturer. However, most energize the pilot solenoid during the cooling mode and deenergize it during the heating mode. This is so that the structure will receive heat during the winter time should some malfunction occur that would prevent the valve from switching positions. These valves instantly reverse the running system refrigerant pressures. They are caused to change positions due to the pressure differential between the high and low sides of the system when they are operating within the rated capacity of the valve. The center capillary tube on the pilot solenoid valve is connected directly into the suction line to the compressor. The other two capillary tubes are connected one to each end of the reversing valve body. When the pilot solenoid is energized the corresponding capillary tube is closed and the other one is opened. This allows the pressure in the valve body corresponding to the closed pilot valve port to be reduced to that of the compressor suction. In the other end the pressure is allowed to build to the compressor discharge pressure. It is this difference in pressure on each end of the sliding piston that causes the valve to switch positions and change the direction of refrigerant flow through the system.

In the following discussion we will consider the pilot valve to be de-energized during the cooling mode; ie. commercial refrigeration.

The refrigerant gas is directed through the main valve body. See Figure 60-2.

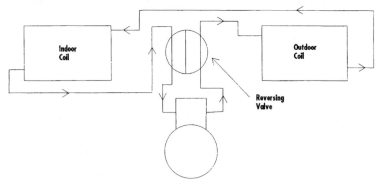

Figure 60-2 De-energized pilot solenoid reversing valve in cooling cycle.

Here, the port on the sliding piston is positioned over two tube openings and it is changing the evaporator and the condenser functions through the operating phases of cooling, and defrosting.

When a defrost cycle is needed, the pilot solenoid coil is energized to change positions of the pilot-valve plunger. See Figure 60-3.

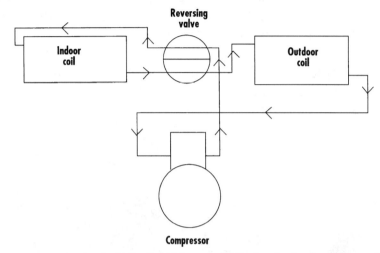

Figure 60-3 Energized pilot solenoid reversing valve in the heating cycle.

The movement of the plunger closes the port on the left with one needle valve and the right pilot port is kept open with the other needle valve. The refrigerant differential that is caused between the two main valves and the chambers due to the action of the pilot valve, the two pistons are caused to immediately move the sliding post. Very soon, both of the valve end chambers have equal pressure in them while the system is operating. However, when the pilot valve is deenergized this condition immediately reverses because of the pressure differential change.

When replacing this type of valve be sure to use the correct replacement or the refrigerant direction may be opposite to that desired. Also, be careful not to overheat the valve body because the sliding pistons are usually made of nylon plastic which will warp out of shape and ruin the valve.

Unit 61: Pilot-Duty Over Loads

Pilot duty overloads are designed for use on low current systems, such as the low voltage control circuit. Their only purpose is to protect the motor from damage due to an overload. Their contacts will not withstand high current flow without damage to them. This type of overload usually has four terminals, two for the high current device and two for the low current circuit. See Figure 61-1.

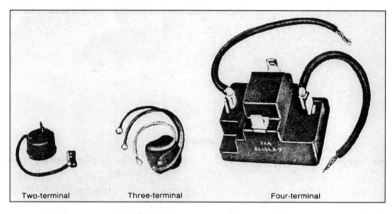

Two-terminal   Three-terminal   Four-terminal

Figure 61-1 Current type pilot duty overloads.
(Photo by Billy C. Langley.)

The current overload is used on systems of almost any size. The high current circuit passes through a heating device that causes the contacts in the low current circuit to open when a certain amount of heat is generated by the flow of current. When the contacts open, the low current circuit is broken. See Figure 61-2.

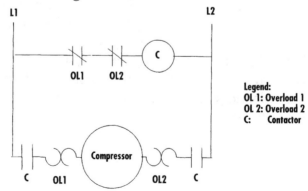

Legend:
OL 1: Overload 1
OL 2: Overload 2
C: Contactor

Figure 61-2 Current type pilot duty overload schematic.

312

OPERATION: When the heavy load is first started the current is high, however, the time delay that is built into the overload does not allow it to trip until a certain amount of heat is generated in the overload. When the motor reaches operating speed the current flow is reduced to the normal operating current. The system will operate until an overloaded condition arises. At this point the heating device will start to heat. When sufficient heat is sensed the overload will open the contacts in the low current circuit. The contactor, relay, or starter coil is deenergized allowing the contacts to open and stop the motor. If the overload is a manual reset type, a service technician or the equipment operator must reset it before the system will operate again. If it is an automatic reset overload, after the heating device has cooled sufficiently the contacts will reset and the system will start automatically. In either case the service technician should be called on to check out the system because there is usually a problem causing the overload to operate.

Currently solid-state type overloads are being used on some equipment. The application dictates the design used on a given piece of equipment. Usually the motor manufacturer places a small sensor in the hot spot of the motor winding which is connected to an external electronic module located in the equipment control panel. When the temperature of the motor winding increases, the resistance of the overload also changes. The electronic module will sense this change in resistance and will open a set of pilot duty contacts in the control circuit. Some manufacturers use multiple resistors in a motor winding to provide a rapid response to an overloaded condition. See Figure 61-3.

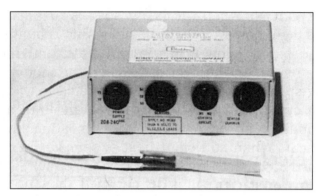

Figure 61-3 Electronic overload with sensor. (Courtesy of Robert Shaw Controls Company.)

TROUBLESHOOTING: Overloads are usually fairly simple devices to test. They have only one set of pilot duty contacts and a heater used to heat the bimetal in the high current part of the overload. An ohmmeter can be used to determine if continuity is indicated through the contacts, the heater coil, and the high current circuit. When continuity is indicated through these components the next step is to check the overload under operation. To do this, place an ammeter on one of the high current wires to the overload. Be sure that a high enough scale is used to protect the ammeter. Start the motor and observe the starting current flow as well as the running current. If the current draw of the motor is within specifications and the overload opens there is a problem with the overload and it must be replaced. Be sure to use an exact replacement so that proper motor protection will provided. If the motor current is high and the overload does not trip and stop the motor, the overload is stuck closed and must be replaced. Be sure to use an exact replacement. If the current draw is high and the overload trips to stop the motor it is probably operating properly. The trouble is with the motor or something is causing the motor to draw excessive current. The problem must be found and corrected.

Unit 62: Temperature Sensing Methods

Temperature sensing is probably the most important step in air conditioning and refrigeration applications. Even though sensing temperature is a single indication there are several different methods used to make these readings. The following discussion will cover some of the most important methods used to sense and measure the temperature of an air conditioned or refrigerated space.

METAL EXPANSION: It is common knowledge that metal when heated will expand and contract when cooled. Each type of metal will expand and contract in different amounts when subjected to a given amount of heat. The expansion and contraction of metal is directly proportional to:

1. The amount of heat applied to it

2. The type of metal used.

Example: If we heat a metal bar, the bar will expand in direct relation to the amount of heat applied to it and the type of metal used. See Figure 62-1.

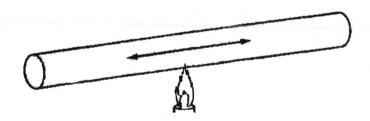

Figure 62-1 Heating a metal bar.

In this case the bar will expand in length in both directions. When the metal is allowed to cool to its original temperature it will contract to its original length. The amount of expansion and contraction will be small, however, we can fasten one end of the bar to a stationary object and when the bar is again heated it will expand in only one direction.

When the bar is used in this manner, it can be made to operate a set of electrical contacts or some other device through a series of linkages and pivots. See Figure 62-2.

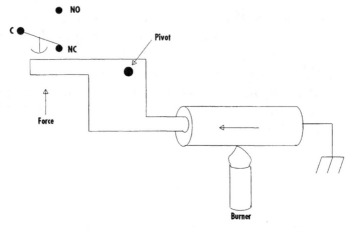

Figure 62-2 Expansion of metal bar operating a set of electrical contacts.

The small amount of expansion becomes large when transferred through the linkages and pivots. When used to operate a set of contacts the linkage must be spring loaded to prevent arcing and burning of the contacts when they open or close the circuit. The spring would be used to provide the necessary snap action to the contacts. The hot wire compressor starting relay is one type of device that makes use of this phenomenon.

Thermometer: The mercury thermometer makes use of the expansion and contraction of a metal. The metal used in this case is mercury. Even though mercury is a liquid at room temperature, it is still considered to be a metal. A small reservoir of mercury is placed at the bottom of a hollow glass tube. When the temperature surrounding the thermometer rises the mercury expands. The only direction it can go is up the hollow tube. It expands in direct relation to the increase in temperature. It will always expand or contract the exact same amount for each degree of temperature change. See Figure 62-3.

Figure 62-3 Mercury Thermometer.

The tube of the thermometer is calibrated and etched to provide a convenient way to determine the current temperature.

BIMETAL: The bimetal also expands and contracts in response to a change in its temperature. The term bimetal is used because they are made from two different types of metal. Each type will have a different expansion and contraction rate. They are bonded together so that they will appear to be only one piece. See Figure 62-4.

Figure 62-4 A Bimetal strip.

When the bimetal is heated one of the pieces of metal will expand faster than the other. This difference in expansion rate will cause the bimetal to warp, or bend. See Figure 62-5.

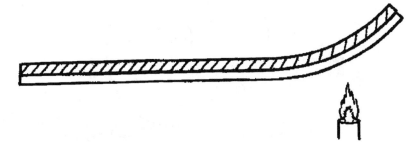

Figure 62-5 A bimetal warping due to temperature change

If one end is stationary. The other end will move more than if both ends were free to move.

Bimetals are often wound into coils, or spirals, and used in thermostats to regulate the temperature inside a space. See Figure 62-6.

Figure 62-6 A bimetal spiral in a thermostat. (Photo by Billy C. Langley.)

This type of bimetal may also be used in thermometers. Most pocket thermometers used by service technicians are of the bimetal type.

The reason that a spiral is used is to provide more movement of the free end of the bimetal. One end is stationary and the other one is free to move in response to a temperature change. The free end may be attached to a needle to indicate the temperature or a set of electrical contacts inside a thermostat. See Figure 62-7.

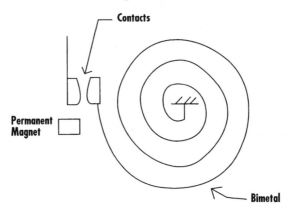

Figure 62-7 A bimetal spiral operating a set of contacts.

Bimetal thermostats use a small permanent magnet to cause the contacts to snap open or closed to prevent them burning. The magnet causes the contacts to snap closed and when the force of the expanding bimetal is strong enough to overcome the strength of the magnet it will cause the contacts to snap open. This is a very inexpensive and popular type of thermostat used for heating, cooling, and refrigeration applications.

THERMOCOUPLE: A thermocouple is made by joining two pieces of different types of metal together at one end (the hot junction)and preventing the other ends from touching (the cold junction). See Figure 62-8.

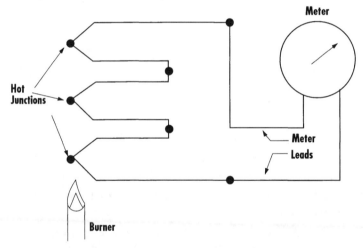

Figure 62-8 A thermocouple used to change heat energy into electrical energy.

When the correct amount of heat is applied to the joint of the two metals a small dc electrical voltage will be generated. The amount of current produced will be proportional to:

1. The two types of metals that are used to make the thermocouple

2. The difference in the temperature between the heated junction and the cold junction.

Thermocouples are used as safety devices in heating equipment, as sensing devices in electronic thermometers, as well as for many other uses when a temperature is to be sensed.

The type of voltage produced will be dc and will be a very small amount. Usually measured in the millivolt range. (One millivolt equals 0.001 volt.) A single thermocouple such as those used in heating equipment will produce about 35 millivolts when heated to approximately 2500 degrees F. When a higher voltage is needed more thermocouples are connected in series. See Figure 62-9.

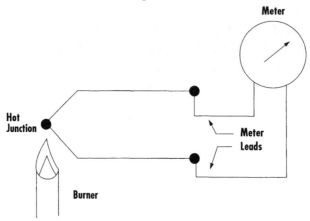

Figure 62-9 Thermocouples connected in series.

The voltage produced by the series connection is increased but it does not add 35 millivolts for each thermocouple added. The resistance in the circuit will reduce the amount produced.

Because the voltage produced by thermocouples is so small it is necessary that all connections be kept tight and cleaned. Any amount of resistance to the flow through the circuit will reduce the amount of voltage reaching the device it is controlling.

RTD (RESISTANCE TEMPERATURE DETECTORS): RTDs are used to accurately measure temperatures ranging from -328° to + 1166°F. They are available in a variety of styles that may be used in different types of applications. See Figure 62-10.

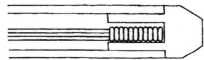

Figure 62-10 RTD in probe form.

The sensing element in an RTD is made from platinum wire. The resistance of platinum wire increases with an increase in temperature and decreases with a drop in temperature. It will change at exactly the same rate for each degree of temperature change at the sensing element.

The RTD makes use of a small coil of platinum wire encased in a copper case. The copper is used to allow the temperature at the element to be sensed faster. The resistance for each degree is shown in Table 62-1. The temperature is in degrees Celsius and the resistance in ohms.

Table 62-1	
A TYPICAL TEMPERATURE	
AND RESISTANCE CHART FOR RTDs	
Degrees C	*Resistance*
0	100
50	119.39
100	138.5
150	157.32
200	175.84
250	194.08
300	212.03
400	247.06
450	264.16
500	280.93
550	297.44
600	313.65

EXPANSION BECAUSE OF PRESSURE: Because of the fact that some chemicals will expand when heated and will contract when cooled allows them to be used to sense a change in temperature. Many of the thermostats used in commercial air conditioning use this principle to control equipment operation.

The sensing devices are made from a very thin brass bellows, a capillary tube, and a sensing bulb all hermetically sealed to prevent leakage of the gas from the sensing system. See Figure 62-11.

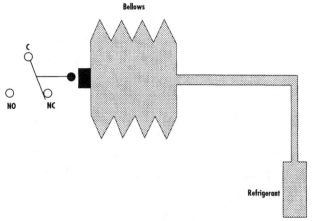

Figure 62-11 A temperature sensing device using pressure inside a sealed system.

The gas used in the enclosed system will have exactly the same amount of expansion or contraction for each degree change in temperature at the sensing element. It acts exactly the same as the refrigerant used in a refrigeration system. When the temperature increases at the sensing element, the pressure increases in the closed system and the bellows will expand and force the switch to either make or break a set of electrical contacts in the circuit. This action will depend on whether the system is heating or cooling. When the temperature at the sensing element drops the pressure inside the closed system will reduce an exact amount and will cause the set of contacts to change position. A temperature sensing device that uses this method is termed a bellows thermostat.

ELECTRONIC THERMISTORS: Thermistors are simply temperature-sensitive semi-conductor devices. That is they are not really good conductors of electricity. Thermistors do not have an exact coefficient of expansion and contraction with a change in temperature at their sensing element. Therefore, they are not usually used as temperature sensing devices. In most applications, they are used as set point devices rather than temperature sensing devices. A set point device may be defined as one that activates some type of process or circuit when the temperature at the sensing element reaches a predetermined point.

Thermistors are very popular as overload devices placed inside the windings of an electric motor. When the temperature of the winding reaches a predetermined point the resistance of the thermistor will increase to a point that a set of contacts in the control circuit will open and stop the motor to prevent it from being damaged.

Thermistors are available in two different types. One has a positive temperature coefficient **(PTC)** and the resistance of the semi-conductor material increases with an increase in temperature.

When the temperature drops the thermistor becomes less resistive to current flow. The other type has a negative temperature coefficient **(NTC)** and the resistance of the semi-conductor material decreases with an increase in temperature. When the temperature drops the resistance of the semi-conductor material increases. The type of operation desired will determine the type to be used in a given situation. The operating range of thermistors is very wide. They are used in applications that have a temperature range of from -100° to +300°F.

When they are used as temperature sensing devices the controlled device must be calibrated to match the type of thermistor being used. When replacing a thermistor be sure to use an exact replacement or the system will not operate as designed.

THE PN JUNCTION DIODE: Before undertaking the study of a P-N junction, let us first consider the condition of the separate P-type and N-type materials before they are joined. See Figure 62-12.

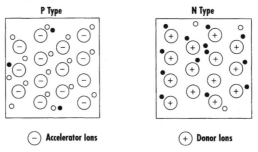

Figure 62-12 P-type and N-type silicon before joining.

The two pieces cannot be merely pushed together. The crystalline structure must remain together at the junction. The P-type material is made up mostly of silicon atoms, a small number of acceptor atoms, holes (or majority carriers), and a few electrons. The acceptor atoms are shown as negative ions because at room temperature all of them will have accepted a valence electron from a neighboring atom to fill its empty covalent bond. These acceptor ions are fixed in the crystal structure of the silicon and cannot move. The number of majority carriers depends on the number of acceptor atoms. The number of electrons (minority carriers) is dependent on the material temperature.

The N-type material is made up mostly of silicon atoms, a small number of donor impurity atoms, free electrons (or majority carriers), and a small number of holes, because of the thermal generation of electron-hole pairs. The silicon atoms not illustrated in the figure and should be thought of as a continuous crystal structure. The positive ions represent donor impurity atoms because at room temperature all of them have released one electron. These donor ions are fixed in the crystalline structure and cannot move. The number of majority atoms is dependent on the number of donor atoms. The number of holes (or minor carriers) is dependent on the material temperature.

When the P and N areas are joined together to form a single crystal structure, there is a high concentration of holes (the majority carrier) in the P region. See Figure 62-13.

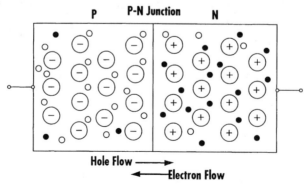

Figure 62-13 P-N crystal formed by joining the P and N regions.

There is a very low concentration of holes in the N region. The holes are the minority carriers in that region. It is the difference in concentration of holes in the semiconductor material that allows the flow of electric current. The holes in the P region will immediately begin diffusing from the area of high hole concentration across the junction into the region having a low concentration of holes. Also, there is a high concentration of free electrons, the majority carrier, in the N region. In the P region, there exists a low concentration of free electrons, the majority carrier. Thus, the electrons are diffused from the N region to the P region across the junction.

As the P region holes diffuse across the junction, they encounter many free electrons. Each diffusing hole combines immediately with a free electron and then disappears. The electrons that are diffusing from the N region across the junction to the P region immediately combine with the available holes and disappear. Not all of the holes from the P region or all of the free electrons from the N region diffuse across the junction. It is only those majority carriers near the junction that diffuse across. This is possible because there is a space-charge region across the junction of the P-N crystal after some of the majority carriers have diffused across the junction and combined. See Figure 62-14.

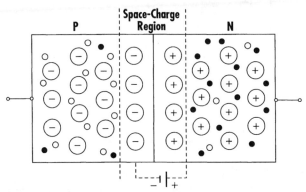

Figure 62-14 P-N crystal after some of the majority carriers have diffused.

Only the fixed ions remain close to the crystal junction

with negative ions on the P side and positive ions on the N side. There are no charge carriers in this region. It is these fixed charges close to the junction, that are caused by the diffusion of major carriers, which have a discouraging effect and eventually prevent any further diffusion. Should a hole from the P side attempt to diffuse across the junction, it is repelled by positively charged donor ions on the N side of the junction. Also, any electrons trying to diffuse across the junction from the N side are repelled by the negatively charged acceptor ions on the P side of the junction.

The space-charge region is a repelling electrical force. It is the fixed charges on the opposite sides of the junction that produce a potential (electrical) barrier, the same that a battery would produce. If we measured the voltage across the space-charge region we would find a potential equal to this barrier. It would be approximately 0.3 V for germanium and 0.7 V for silicon. See Figure 62-14. On either side of the junction away from the space-charge region, the charges are concentrated exactly as they were before joining the P-N regions. The charges, both positive and negative, are equally distributed. Thus, the net density of the charge is almost zero in all areas except the space-charge region. See Figure 62-15.

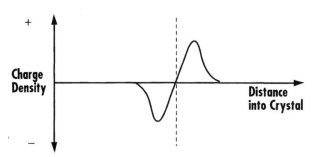

Figure 62-15 Charge profile of a P-N crystal.

In the space-charge region there exists a large positive charge on the N side of the junction and a large negative charge on the P side of the junction.

The P-N junction, because of the space-charge region, offers a resistance to any majority carriers attempting to diffuse across the junction. Thus, for a free electron in the N region to cross the junction into the P region, it must gain enough energy to overcome this barrier. Also, for a hole in the P region to cross over into the N region a valence electron in the N region must overcome the barrier to fill this hole and have the hole appear on the N side. Because of the heat energy gained at room temperature, some of the majority carriers will overcome the barrier and diffuse across the junction. Some of the holes in the region and a few of the electrons in the N region will gain enough energy to overcome the resistance barrier and diffuse across the junction. This small amount of diffusion results in a majority-carrier diffusion across the P-N junction. This diffusion current is the sum of the hole current plus the electron current. In contrast, this diffused current is exactly balanced by a current of equal value set up in the opposite direction and formed by the minority carriers. It is the attraction of the space-charge region that causes this minority-carrier current. See Figure 62-14. Should an electron-hole pair be produced by the heat energy in the space-charge region, the electron will be attracted toward the N region by the positively charged ions and the hole will be attracted toward the P region by the negatively charged ions. This attraction causes a minority-carrier drift current across the P-N junction in the direction opposite to the flow of the majority-carrier diffusion current. When there are no external power connections on the P-N crystal, the majority-carrier diffusion current and the minority-diffusion current cancel each other out, resulting in a net junction current of zero. See Figure 62-16.

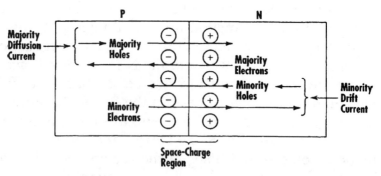

Figure 62-16 Currents across a P-N junction with no external connections.

When an external power source is applied to a P-N junction, it is said to be biased and the voltage applied is said to be bias voltage. When no external power source is applied to the P-N junction, the junction is unbiased.

Temperature Sensing with A P-N Junction (Diode): The diode is becoming popular for temperature sensing because it is very accurate and is linear in operation. That is they perform exactly the same every time for each degree change in temperature.

When diodes are used as temperature sensors, a constant current passes through the diode. See Figure 62-17.

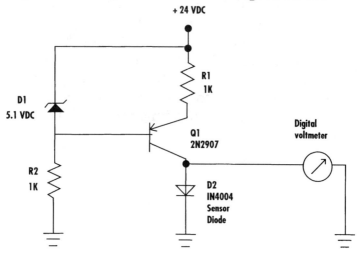

Figure 62-17 Current regulated by a diode.

In this type of circuit the resistor R2 regulates the current flowing through both the transistor and the sensing diode. The ohmic value of R2 also determines the amount of the current that will flow through the diode. The R1 diode is a 5.1-volt sensing diode which is used to provide a constant voltage between the emitter and the base of the PNP type transistor. Resistor R1 is used to limit the amount of current that can flow through the zener diode to the transistor base. The D2 diode is a common silicon diode that is being used as a temperature sensor in this circuit. A digital voltmeter connected across the diode terminals will indicate a voltage drop between 0.8 and 0 volts. The temperature of the diode is what determines the amount of voltage across its terminals.

When the temperature of the diode drops the voltage drop across it will be increased. When the temperature of the diode is increased the voltage drop across it will be decreased. This type of diode has a negative temperature coefficient.

Unit 63: Manual Switches

The manual electrical switch is used to either connect or disconnect a source of power to an electrical circuit. The switch is operated manually rather than automatically. In most instances they are used when automatic control is not necessary or desirable. This type of switch may be a knife type or a push-button type depending on the needs of the system being controlled. They may be of the single-pole, single-throw; single-pole, double-throw; double-pole, single-throw; double-pole, double-throw type. See Figure 63-1.

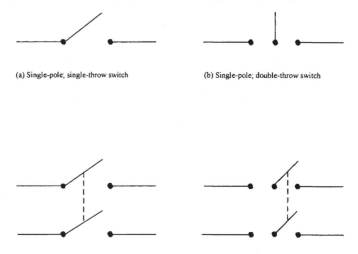

(a) Single-pole; single-throw switch   (b) Single-pole; double-throw switch

Figure 63-1 Manual Switch symbols.

The term poles indicates the number of contacts included in the switch. The term throw indicates how the switch may be operated for the desired purpose. For instance, a single-pole, single-throw switch has two positions and one set of contacts. The positions are either open or closed depending on the equipment operation. The double-pole, double-throw type has three positions; closed to one circuit and open to another; open; and closed to another circuit and open to the other. This type of switch will close two sets of contacts when it is in the closed position to either circuit. These switches are generally used to control more than one circuit. Refer to Figure 63-1.

DISCONNECT SWITCHES: This type of switch is usually used as a manual disconnect for a circuit that needs to be shutdown for service or repairs. They may be equipped with fuses or they may be simply single-throw switches with the necessary number of poles for the application. See Figure 63-2.

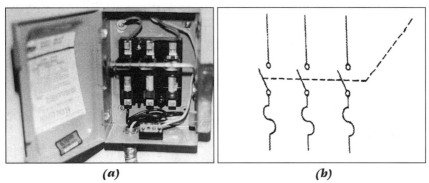

(a)   *(b)*

Figure 63-2 Disconnect switch. (Photo (a) and diagram (b) by Billy C. Langley.)

Their major use in air conditioning and refrigeration systems is to allow the service technician to shutdown the equipment so that it can be serviced or repaired. Some electrical codes require that the disconnect switch be located within five (5) feet of the unit being serviced. When used in this manner, the switch is a safety device to protect the technician from electrical shock or other injury.

Unit 64: Push-Button Station Switches

Push-button switches are generally used to open or close a circuit by simply pushing the correct button. They are usually used to operate lifts for moving and lifting heavy equipment around the shop. They are sometimes used for testing the operation of equipment that has been repaired, or for any other practical use around the shop. See Figure 64-1.

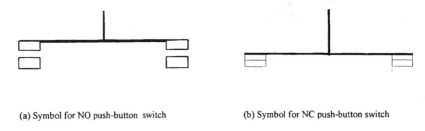

(a) Symbol for NO push-button switch

(b) Symbol for NC push-button switch

Figure 64-1 Push button contact symbols.

Unit 65: Troubleshooting Electric Control Devices

INTRODUCTION: Troubleshooting air conditioning and refrigeration systems generally involves a problem that the user is currently having with equipment operation. About 80 to 90% of the problems with this type of equipment is electrical in nature. The service technician is expected to locate the cause of the problem and correct it or advise the user why it cannot be repaired. Oftentimes it may be difficult to locate the problem with the system. However, if the technician takes the time to talk with the user and ask the right questions the user will at least indicate where to start looking for the problem. Remember, the user is around the equipment every day and knows what it sounds like when it is operating properly. Most users of equipment do not arbitrarily call a service technician just to have one present. There is a problem or the user would not have called.

Of course, the technician must be familiar with the operation of the entire system in order to decipher the chaff from the wheat that the customer is relating to them. When the technician understands the operation of the equipment and its controls, the problem is much more easily and economically located and corrected. Troubleshooting requires that good and accurate instruments be used or any readings may be faulty and lead the technician in the wrong direction. Also, the technician must be able to properly use the tools and instruments to get the most accurate use from them. We will discuss some basic guidelines on testing and servicing air conditioning and refrigeration equipment. The technician should develop a method of troubleshooting that best suits his/her own procedures. Oftentimes the troubleshooting method is a combination of what others use and is modified to suit the technician.

STARTERS AND CONTACTORS: These devices are used to start compressors and other heavy current using devices in the system. They operate alike except that the starter has overloads included in the basic control.

STARTERS: Problems with starters usually are due to four different sources: The contacts, the holding coil, the mechanical linkage, and the overload or overloads.

CONTACTS: The purpose of the contacts is to complete the electrical circuit to a motor or some other electrical device so that power can be supplied to it. During use the contact surfaces become pitted and rough and will not make good contact with the mating part of the contact. When this happens, the motor will draw excessive current and it will sometimes cause the motor to stop on the overload. Contacts can be visually checked for pitting and burning. When pitted contacts are found they should be replaced.

It is not generally a good idea to file the contacts smooth because the chances of them fitting properly are very unlikely. Also, the silver coating on contact surfaces that help in conducting the current through them and provides a harder surface to help prevent pitting of the contacts, is very thin and filing will remove the coating and allow the contacts to pitt much more easily. Only file the contacts when it is necessary to wait for the proper new ones to arrive.

If the contacts are suspected to be the problem but visual inspection does not indicate they are the problem, they can also be checked with a voltmeter. To make this test, connect a lead from the voltmeter on each side of the contact and energize the starter. If there is a reading of 5% or more of the line voltage when the starter is pulled-in and the motor is running, the contacts are pitted and preventing all the current from going to the motor. The contacts must be replaced. Also, the amperage can be checked while the starter is energized. If there is excessive current draw through that part, or parts, of the circuit the contacts are bad and must be replaced.

Sometimes a resistance check of the contacts while they are closed will indicate their condition. The starter must be energized, but the electrical power to the motor must be turned off and the wires disconnected. When the resistance is more than one ohm the contacts are faulty and must be replaced.

Before replacing the contacts in a starter check their alignment with each other. They must be properly aligned or they will not last very long. All sets of contacts must close at the same time to prevent future pitting. They must close with a firm pressure caused by the springs located underneath the movable contact.

Overloads: The overloads on starters are to provide motor protection from an overloaded condition caused by low voltage, high voltage, or grounded or shorted windings in the motor, or a mechanical condition. An overload that occasionally trips is probably not faulty but is doing what is was designed to do. The simplest way to check and overload is to start the motor and check the running amps. If the current flow is normal and the overload continues to trip it is faulty and must be replaced to solve the problem. Be sure to use an exact replacement or the motor may not be properly protected. If the current draw is higher than indicated on the motor nameplate the problem is not with the overload but with something else. The problem must be found and solved.

Holding Coil: The purpose of the holding coil is to pull-in the starter or contactor linkage and close the contacts when the control circuit demands equipment operation. There are only three tests that can be done on a holding coil: Test for open, grounded, or shorted coil.

When the holding coil is suspected of causing the problem make the tests as follows:

To test for an open holding coil, turn off the electricity to the coil, remove one of the coil electrical leads and check for continuity with an ohmmeter. If infinite continuity is indicated the coil is open and must be replaced. Be sure to use an exact replacement or the coil will probably burn-out again in a short period of time.

To test for a grounded holding coil, turn off the electricity to the coil terminals and remove one of the coil wires. Test from each terminal to the frame of the starter or to some other known ground. If continuity is indicated, the coil has grounded and must be replaced. Be sure to use an exact replacement.

To check for a shorted holding coil, turn off the electricity to the coil terminals. Remove one of the wires from the terminal and check the coil continuity with an ohmmeter. If the test indicates a lower than normal continuity, the coil has shorted and must be replaced. Be sure to use an exact replacement coil.

Linkage: The linkage in a starter is what changes the magnetic energy into movement to close the contacts. Sometimes this linkage becomes worn or corroded causing the starter to malfunction. Sometimes a lubricant such as WD 40 or some other type may be sprayed onto the linkage and loosen it so that it will operate. When lubrication does not properly solve the problem the starter must be replaced. Sometimes it is a good idea to also replace the overloads during the process.

CONTACTORS: The problems that occur with contactors are: bad holding coil, bad contacts, bad linkage.

Holding Coil: A holding coil is usually shorted, opened, or grounded. To check for a shorted condition; disconnect the electrical power from the holding coil connections and remove one of the wires from the coil terminal. Set the ohmmeter on ohms R x 100. Touch each of the meter leads to the terminals of the holding coil. If the wrong reading is indicated, the coil is shorted and must be replaced. To check for an open coil; disconnect the electrical power from the holding coil connections and remove one of the wires from the coil terminals. Set the ohmmeter on ohms R x 100. Touch the two meter leads to the coil terminals. If no reading is indicated the coil is open and must be replaced.

To test for a grounded holding coil; disconnect the electrical power from the holding coil terminals and remove one of the wires from one of the terminals. Set the ohmmeter on ohms R x 100. Touch one of the meter leads to one of the coil terminals and the other lead to the contactor frame. If continuity is indicated with these connections the coil is grounded and must be replaced.

When replacing the coil make certain that the voltage rating for the replacement coil is exactly the same as the one being replaced. Also, make certain that the proper bushings and spacers are replaced in their proper places and that the coil is installed with the top and bottom in their correct places to prevent the coil from becoming overheated and possible early failure.

Linkage: The contactor linkage is the components that move the contacts in response to the demands of the holding coil. They consist of various pivots and points that cause the contacts to move as they were designed. When the linkage becomes rusted or corroded the contacts will not be opened or closed properly. In some cases this may be indicated by a humming or buzzing noise at the contactor. When this condition occurs it may sometimes be eliminated or helped by use of a spray such as WD 40 or some other type of lubricant-rust remover. If the contactor still does not operate properly after working it a few times it must be replaced with one having at least the same amperage rating as the one being replaced. If one is used that is much too large for the application the cost will be excessive. If one is used that is too small the contactor will not last and must be replaced sooner than normal.

Contacts: The contactor contacts are what allows the electricity to flow to the motor terminals. When they become worn or pitted they will provide a resistance to the flow of the electricity resulting in a lower than normal current to reach the motor terminals. This will usually result in higher than normal current draw and a motor running at a higher than normal temperature.

There are several ways to check for worn or pitted contacts. Probably the easiest is to remove the electrical power to the contacts. Remove one of the wires to the contact terminals. Set the ohmmeter on R x 1 and be sure to zero the ohmmeter. Allow the coil to pull-in the contactor as it would normally. This can be done by allowing the control voltage to remain on during the test. Touch one of the meter leads to each terminal of the contact. If one ohm resistance is indicated the contacts are faulty and must be replaced. Repeat this test with all the contacts in the contactor. Usually when one set needs replacement all the others do also. Be sure to use the correct replacement contacts with the same current rating as the ones being replaced. In some cases the contacts cannot be replaced and the complete contactor must be replaced. Be sure to use the correct replacement or problems in mounting the new contactor may be encountered.

Another way to check the condition of the contactor contacts is to start the motor and check the voltage across each set of contacts. Be sure to check across the same contact or a improper reading will be indicated. There should be no voltage indicated if the contacts are good. When they are burned, pitted, or out of adjustment there should be some voltage indicated on the meter. When a voltage is indicated the contacts are not mating properly and must be replaced. When replacing the contacts be sure to use the exact replacement for proper operation. In some instances the contacts cannot be replaced. Then the complete contactor must be replaced. Be sure to use an exact replacement or mounting problems may be encountered.

RELAYS: The operation of a relay is very much like the starter and contactor. The contacts in a relay are generally not accessible to the technician without dismantling the relay. However, the contacts can be checked with an ohmmeter to determine their condition. Relay contacts are not designed to carry the heavy current that those used in starters and contactors are. The major problem is determining which terminals go to each set of contacts. There is usually a diagram on the unit or the relay that will provide this information. To test relay contacts first disconnect the electric power. Determine which terminals are a part of which contact set. Set the ohmmeter on R x 1 and check the continuity through the contact set. If continuity is indicated on NO contacts or if there is no continuity indicated on NC contacts the relay is bad and must be relaced. If the continuity test indicates that the contacts are as they are supposed to be, then energize the relay coil and listen for a click which indicates that the relay has pull in. Recheck the contact continuity to determine if the contacts switched position. If they did not the contacts are stuck in the deenergized position and the relay must be replaced. The relay contacts must never be subjected to a voltage or current higher than their rating. To do so will cause immediate failure of the contacts and require replacement of the complete relay.

The relay coil is tested the same way that the holding coil in a starter or contactor is tested. Place the ohmmeter on R x 100. Disconnect the electricity from the coil terminals and remove one of the terminal wires. Place one lead of the ohmmeter on one terminal and the other lead on the other terminal. If continuity is indicated the coil is usually good. If a very low resistance is indicated the coil is shorted and must be replaced. If there is no continuity indicated the coil is open. Either of these last two tests will usually be accompanied by a burnt place on the coil or a burnt electrical odor. It is usually best to replace the complete relay because the coils and contacts are not usually available and the time required to make the repairs would cost more than the relay.

The mechanical linkage of a relay usually gives very little trouble because they are usually more protected from the elements than starters and contactors. Usually sticking contacts will cause more problems with the linkage than the linkage itself.

Section 5: Electrical Control Circuits

INTRODUCTION:

The control circuits for refrigeration, air conditioning, and heating systems have become so complicated, the technicians working in this field must be able to read and understand all types of wiring diagrams in order to successfully do their job. When wiring diagrams are properly read and understood, there is a wealth of information contained in them. When installing a system the installation technician depends on the wiring diagram for information about making the proper connections so the system will operate as designed. Service technicians use the wiring diagram to aid in making troubles diagnosis of the faulty system.

When wiring diagrams are made symbols are generally used because to use pictures of the actual components would be so large and cumbersome that it would be practically worthless to the technician. Each component in the circuit has an almost universally accepted symbol to indicate how it is connected into the system. The technician must be familiar with these symbols, or at least know where to find those with which are unfamiliar. Not all manufacturers use identical symbols for every component, but knowledge of the basic symbols is needed.

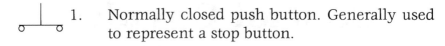

BASIC WIRING SYMBOLS

1. Normally closed push button. Generally used to represent a stop button.

2. Normally open push button. Generally used to indicate a start button.

3. Double acting push button. Has both normally closed and normally open contacts in one push button.

4. Double-acting push button shown different, but means the same thing.

5. Double-acting push button. The dashed line shows that mechanical interlock is used. Thus, when one button is pushed, the other one also moves.

6. Single-pole, single-throw switch (SPST).

7. Single-pole, double-throw switch (SPDT). Notice that this switch has only one pole, the switch arm, but it has two stationary contacts. In the diagram, the switch arm makes contact with the upper stationary contact. When the switch is thrown, contact is made between the switch arm and the lower stationary contact. The switch arm or pole of the switch is generally called the common because it can make contact with either of the two stationary contacts.

8. Double-pole, single-throw (DPST) switch. The dashed line indicates a mechanical interlock between the two arms.

BASIC WIRING SYMBOLS (Cont.)

 9.   Double-pole, double-throw (DPDT) switch.

10.   Off-Automatic-Manual control switch. Basically, this is a single-pole, double-throw switch with a center off position.

11.   Normally open (NO) set of contacts.

12.   Normally closed (NC) set of contacts.

13.   Fuse.

14.   Fuse.

15.   Transformer.

16.   Coil.

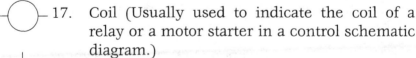 17.   Coil (Usually used to indicate the coil of a relay or a motor starter in a control schematic diagram.)

18.   Pilot lamp of light.

19.   Lamp or light bulb.

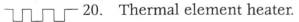 20.   Thermal element heater.

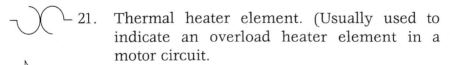 21.   Thermal heater element. (Usually used to indicate an overload heater element in a motor circuit.

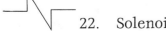 22.   Solenoid coil.

BASIC WIRING SYMBOLS (Cont.)

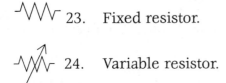 23.   Fixed resistor.

24.   Variable resistor.

25.   Variable resistor.

26.   Single-pole circuit breaker.

27.   Double-pole circuit breaker.

28.   Capacitor. Either run or start.

29.   Normally closed (NC) float switch.

30.   Normally open (NO) float switch.

31.   Normally closed (NC) pressure switch.

32.   Normally open (NO) pressure switch.

33.   Normally closed (NC) temperature switch. (Normally closed thermostat.)

34.   Normal open (NO) temperature switch.

Electrical Applications for Air Conditioning & Refrigeration Systems

BASIC WIRING SYMBOLS (Cont.)

35. Normally closed (NC) flow switch. This symbol is used for both fluid and air sensing.

36. Normally open (NO) flow switch.

37. Normally closed (NC) limit switch.

38. Normally open (NO) limit switch.

39. Normally closed (NC) On-delay timer contact. Usually shown as DOE in schematic diagrams. Used to show a Delay-On Energize.

40. Normally open (NO) On-Delay timer contact.

41. Normally closed (NC) Off-Delay timer contact. Usually shown as DODE in schematics. This indicates a Delay on De-energize.

42. Normally open (NO) Off-Delay timer contact.

43. DCV battery.

44. Electrical ground.

45. Mechanical ground.

46. Crossing wires with no connection.

47. Crossing wires with no connection.

48. Wires crossing with an intersection. The dot on the cross point is called a node. It indicates an electrical connection.

49. Rotary switch

Unit 66: Wiring Diagrams And Schematics

Wiring diagrams are generally considered to be pictorial diagrams which show the individual components in their relative location in the control circuit.

A schematic diagram, however, uses symbols to designate the components. These diagrams range from simple to very complex, depending on the purpose of the diagram. Schematic diagrams are the most popular because they tell more about the operation of the unit as well as they are smaller and require less space for the diagram. Most residential systems are equipped with a simple schematic diagrams. Commercial systems have more complex schematics because of the equipment requires more control. The layout method of both the commercial and residential schematics are usually the same but the different requirements make them more complex.

SIMPLE SCHEMATIC: The power source shown in a schematic diagram is two lines usually drawn downward. They are labeled as L1 and L2. The voltage requirements for this circuit will be provided through these wires. The lines are connected to a device which uses the electricity for operation. When the circuit is closed the electrons will flow though the circuit and cause the desired operation. See Figure 66-1.

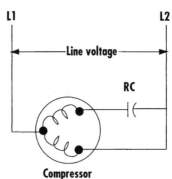

Figure 66-1 Schematic of a complete split-phase circuit.

All of the circuit loads are connected between L1 and L2. The switches that control operation of the loads are also included in the schematic. See Figure 66-2.

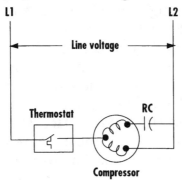

Figure 66-2 Schematic of a circuit with a thermostat.

This schematic shows a compressor motor and an operating control (thermostat in this drawing). When the thermostat demands cooling the circuit to the compressor motor is completed and the compressor runs. When the thermostat is satisfied it opens the circuit to the compressor motor and the cooling process stops. The electric power is supplied through wires L1 and L2, through the disconnect switch and to the thermostat. The compressor motor is the device that uses the electricity. Switches do not use electricity. They simply open or close the circuit.

The equipment manufacturer includes complete schematic diagrams with the unit when it is shipped. They are generally pasted to one or more of the electric control panel covers. They should be consulted when a operation problem occurs with the unit. See Figure 66-3.

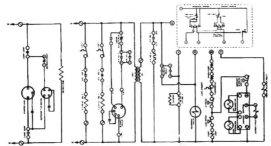

Figure 66-3 Complete system schematic for a Lennox heat pump system. (Courtesy of Lennox Industries.)

Usually a legend is included so that references to the components can be more easily made. Sometimes the complete schematic diagram is not on one piece of the equipment. The condensing unit schematic is in the condensing unit and the indoor section schematic is in the indoor section of the unit. The legend contains both the letter designations and the component name used on the equipment. A wiring schematic is a valuable tool when troubleshooting a system having electrical problems.

DRAWING A SCHEMATIC WIRING DIAGRAM:

There is one thing that must be remembered when drawing a wiring diagram. It is that every component has its own circuit which is separate from any other circuits in the overall electrical system. It is usually much easier and simpler to draw one circuit at a time to avoid confusion.

The first step is to determine where and how the line voltage is going to get to the unit. In most installations this will be from a fuse box or a circuit panel. Remember that the component must be connected to both sides of the line voltage to operate.

Lets start with a compressor that has line voltage connected directly to it through a fused switch and L_1. The electricity passes through the compressor motor winding and returns to the other side of the line L_2. See Figure 66-4.

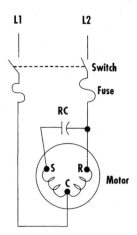

Figure 66-4 Motor connected through a fused switch.

When the switch is closed and the fuses are good the motor will run. It will continue to run until the fused switch is opened. The next step is to add a control to the circuit so that the motor will operate automatically without someone being there to operate the switch. See Figure 66-5.

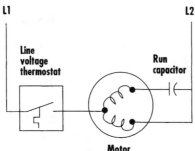

Figure 66-5 Motor circuit with a line voltage thermostat.

The thermostat may be placed in either line, but, it is usually installed in the "hot" line of a 120 volt ac system. When 240 volts are used it does not matter which line it is placed in. Now the motor will run when the thermostat demands cooling.

To make the system more efficient and more effectively do what it was designed to do we can add other components; such as a running capacitor to the circuit. See Figure 66-6.

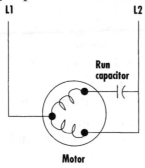

Figure 66-6 A circuit auxiliary winding and run capacitor.

Other items that can be installed for motor protection and compressor protection are overloads, high-pressure, and low-pressure controls. These are installed in the line to the common terminal of the compressor. See Figure 66-7.

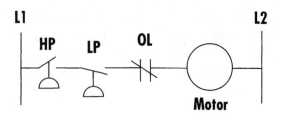

Figure 66-7 Circuit containing high and low-pressure controls and overload.

This circuit is now complete for a PSC compressor motor.

We can now draw the diagram for the condenser fan motor. It is also a PSC motor. The electric wires are connected to the outlet terminals of the switch, just as those to the compressor were connected. See Figure 66-8.

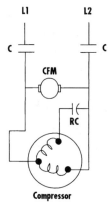

Figure 66-8 Connecting condenser fan circuit

Remember that one side of the line connects to the run terminal of the motor and the other connects to the common terminal of the motor. A run capacitor is also connected between the run and start terminals of the motor. An overload is also used in the line to the motor common terminal.

All other circuits may be completed in the same manner for other compressor motors, fan motors, and the indoor components.

A transformer may also be added that will reduce the

line voltage to 24 vac for the control circuit. This is desirable because of safety, lower installation cost, and more accurate temperature control. See Figure 66-9.

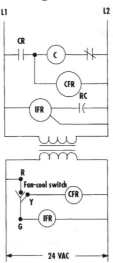

Figure 66-9 Cooling system using a transformer

The purpose of the transformer is to provide low power to the control components that control the operation of the line voltage components. Such as; the compressor contactor **(C)**, and the indoor fan relay (IFR).

READING SCHEMATIC DIAGRAMS: There are a few rules, or procedures, that must be learned before you can become efficient in reading schematic diagrams. It is usually best to memorize these rules so that they can be recalled immediately when reading the schematic.

1. Schematic wiring diagrams are read from left to right and from top to bottom. Just like reading a book.

2. The electrical symbols of the components are always shown in their de-energized, or off, position.

3. Relay coils and their contacts are shown using the same numbers or letters. The coil will be placed in that part of the circuit which controls is action and the contacts will be located in that part of the circuit that they control. Thus, the coil and its related contacts may be separated by some distance.

4. When the coil is energized all of its contacts change position. That is, the closed contacts will open and the open contacts will close.

5. Components that are used to cause the system to stop are shown normally closed. They may be connected in electrical series. Figure 66-10 is an illustration of this principle.

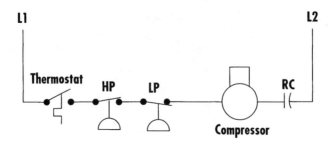

***Figure 66-10** Components that are designed to cause operation of the equipment to stop are normally closed (NC) and are connected in series.*

Both of the switches are normally closed and connected in series. When wired in this way either switch can stop operation of the motor.

6. Components that provide a start function are wired in electrical parallel and shown normally open. See Figure 66-11.

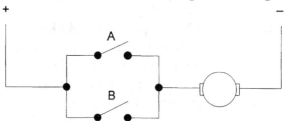

***Figure 66-11** Components used to provide a start function are wired in parallel and are shown normally open (NO).*

In this diagram, when either of the switches is closed the circuit to the motor will be completed causing it will run.

7. There must be a completed circuit before the electric current will flow through the components.

We will discuss a simple control circuit that is used to manually control the operation of an electric motor. See Figure 66-12.

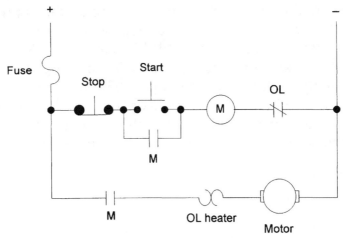

Figure 66-12 Simple start-stop push-button control circuit.

In this schematic, both the control circuit and the motor circuit are together.

In this schematic there is not a complete circuit to C, the motor contactor coil because a start-stop contact is open causing the C_1 and C_2 auxiliary contacts to be open. The circuit is not completed to the motor because the set of contacts C_2 in that circuit are also open. Contacts C_1 are also open. These contacts are very small and are to be used as pilot duty contacts (low current) only. Contacts C_1 are usually referred to as holding contacts. They are closed when the start button is pressed because the contactor coil C is energized which closes the contacts. See Figure 66-13.

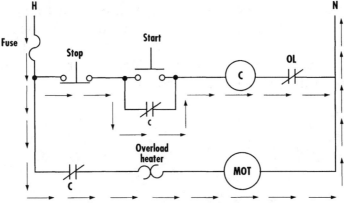

Figure 66-13 Current flow when holding coil is energized.

They will remain closed until the stop button is pressed. When the start button is pressed, both sets of contacts close and remain remain closed until the stop button is pressed. Then the contactor holding coil C is de-energized which allows both sets of contacts to open. The motor stops operating and the circuit to the contactor coil C is also opened by the holding contacts. The circuit will remain in this position until the start button is again pressed. See Figure 66-14.

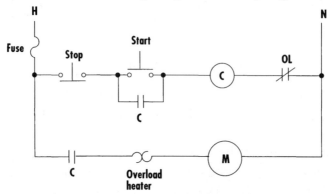

Figure 66-14 Stopping the motor with the stop button.

The motor overload contacts are wired in series with the contactor coil C. Should the motor become overloaded for some reason, the overload heater (OLH) will cause the OL contacts to open, de-energizing the contactor coil. Then both sets of C contacts open to stop the motor and de-energize the control circuit. The motor cannot be restarted until the overload heater has cooled, or has been replaced, the overload contacts are closed again, and the start button pressed again.

A typical schematic wiring diagram for a heating and cooling unit is shown in Figure 66-15.

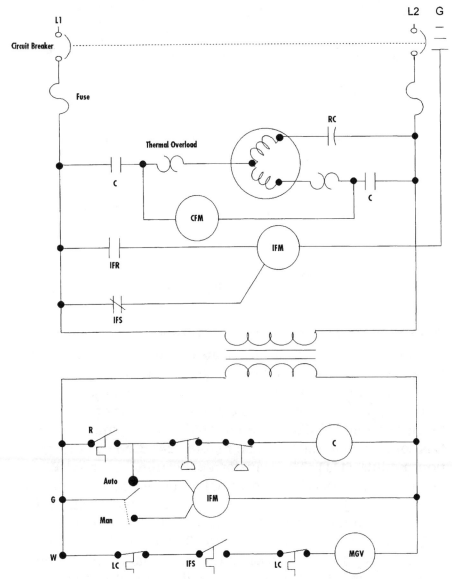

Figure 66-15 Typical schematic wiring diagram for heating and cooling.

READING A HEATING UNIT SCHEMATIC DIAGRAM:

We will discuss a simple heating only schematic wiring diagram to illustrate how the procedure is done. Some service technicians will use the following methods and others will use a combination of methods to determine the operation of a system. See Figure 66-16.

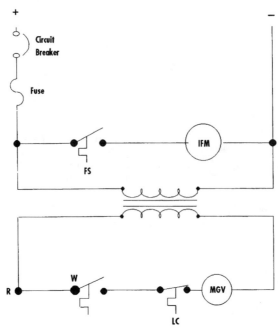

Figure 66-16 Simple Schematic wiring diagram of a heating unit.

The electric power to the heating unit is fed in through L_1 and L_2. These lines are connected to the primary side of the transformer. When power is supplied to the transformer primary power is also available at the terminals of the transformer secondary. The terminal on the transformer marked C is connected to the terminal marked 24V GND on the ignition module **(IM)**. The transformer terminal marked R, or Load, is connected to one terminal of the limit control **(LC)**. This is usually a red wire. The other terminal of the limit control is connected to the R terminal of the thermostat **(T)**. This is also usually a red wire. The H or W terminal of the thermostat is connected to one terminal of the air providing switch **(APS)** and to one coil terminal of the combustion air blower relay **(CABR)**. The other terminal of the APS is connected to the terminal marked 24V on the ignition module.The remaining terminal of the combustion air blower relay is connected to the transformer C terminal.

The pilot burner/ignitor sensor **(PBIS)** is connected to the high tension terminal of the ignition module. This wire carries very high voltage and must be protected from damage for proper operation of the system.

The terminal MV on the ignition module is connected to the main valve terminal of the 2nd operator of the dual valve combination gas control **(DVCGC)**. The terminal MV/PV on the ignition module is connected to the common terminal **(C)** of the dual valve combination gas control. Terminal PV on the ignition module is connected to the pilot valve **(PV)** terminal on the dual valve combination gas control. The GND terminal of the ignition module is connected to the pilot burner for a ground connection to the unit. This ground connection must be a solid connection or the unit will not operate properly.

There is also a temperature operated fan control **(FC)** which has normally open contacts. The fan control and the circulating fan motor are connected directly across the line voltage to the unit.

DETERMINING UNIT OPERATION FROM SCHEMATIC DIAGRAM: Power is delivered to the unit through wires L_1 and L_2 to the primary side of the transformer **(TR)** Unless the transformer is bad there will also be 24V power at the secondary terminals of the transformer. Power will flow through the red wire to the limit control **(LC)**, which has normally closed contacts, and to the red terminal of the thermostat **(T)**. When the thermostat demands heat its contacts will close, completing the circuit to the air proving switch **(APS)**, the combustion air blower relay **(CABR)**, and the heat anticipator. The heat anticipator is a resistor placed in electrical series with the remainder of the control circuit. Its purpose is to provide a false heat inside the thermostat to cause it to stop the main burners before the desired temperature is reached inside the space to prevent system override. It is adjustable so that it can be set to the actual current draw of the heating circuit. The combustion air blower relay contacts, which are normally open, close to start the combustion air blower **(CAB)**. When the volume of air that the equipment manufacturer requires is detected the air proving switch contacts, which are normally open, close and complete the circuit to the 24V terminal on the ignition module **(IM)**. The terminals MV/PV and PV on the ignition module energize the pilot valve 1st operator to allow gas to flow to the pilot burner. Also at the same time the pilot burner/ignitor sensor is provided with the voltage required to light the pilot gas. When a flame is proven the circuit through the pilot burner/ignitor sensor, the pilot burner ground and back to the GND (burner) terminal on the ignition module terminal. MV on the ignition module is energized which energizes the coil terminal of the 2nd operator of the dual valve combination gas control. The dual valve combination gas control will open and allow gas to flow to the main burners.

When the heat exchanger has heated sufficiently for the fan control contacts to close, which is usually 135°F to 150°F, the circulating fan motor will be energized and start blowing heated air into the space.

The system will remain in this mode until the thermostat is satisfied and de-energizes the dual valve combination gas control to stop the flow of gas to the furnace. When the thermostat is satisfied the coil in the combustion air blower relay is also de-energized. When the required air volume drops below that required for combustion the air proving switch contacts open. The blower will continue to operate until the heat exchanger has cooled sufficiently to cause the fan control contacts to open and stop the circulating air blower. The contacts usually open at about 100°F.

The heating unit will remain in this position until the thermostat again demands heat. All of the controls are now in their de-energized or normal position.

A SIMPLE SCHEMATIC WIRING DIAGRAM FOR A COOLING UNIT: This type of schematic is just as simple as the one we just discussed. The only difference is the components and their sequence of operation. However, the operation of a cooling unit can be just as easily determined from a schematic diagram. Some manufacturers use a color code to help in following the line voltage portion of their diagram. See Figure 66-17.

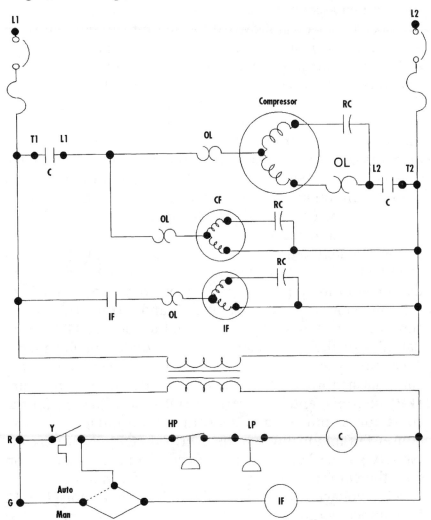

Figure 66-17 Simple schematic of a cooling unit.

The control portion of the diagram should always use the somewhat standardized color code.

Electric power is provided to the unit through wires, L_1 and L_2. There is a manually operated switch provided for safety and control of the line voltage to the unit. Usually the line voltage components are shown on the top part of the diagram and the control or low voltage is shown on the bottom part.

Electrical power circuits are always protected by circuit breakers or fuses. Depending on the desires of the installation personnel. Fuses may be placed in the disconnected switch mentioned above. Circuit breakers are usually placed inside an enclosure designed to hold them. The electric wires are connected from the load terminals inside the disconnect switch to the line side of the contactor **(C)** contacts terminals L_1 and L_2. The compressor **(COMP)** and the condenser fan motor **(CFR)** are connected to the load side of the contactor contacts, terminals T_1 and terminal T_2 on the contactor. Terminal T_1 is connected to the run **(R)** terminal of both the compressor and the condenser fan motor. Terminal T_2 is connected to the common terminal of the compressor through high pressure **(HP)** and low pressure **(LP)** controls, and the overload **(OL)**. Terminal T_2 on the contactor is connected to the common terminal of the condenser fan motor through the overload **(OL)**. A run capacitor **(RC)** is connected between the R and C terminals on both the compressor and the condenser fan motor.

Line voltage is supplied to the indoor fan motor **(IF)**. L_2 is connected directly to one terminal of the indoor fan motor and L_1 is connected to one terminal on the indoor fan relay **(IFR)** which is operated by a coil located in the control voltage circuit. The other terminal of the indoor fan relay is connected to the other terminal of the indoor fan motor.

L_1 is connected to one side of the transformer **(TR)** primary and L_2 is connected to the other transformer primary connection. When the disconnect switch is closed power is fed directly to the transformer primary to provide low voltage **(24V)** to the control voltage circuit.

The R or load terminal on the transformer secondary is connected to the R terminal on the indoor thermostat with a red wire.

A green wire is connected from the G terminal on the thermostat to one side of the coil of the indoor fan relay. The other connection on the coil is connected to the C terminal on the transformer secondary.

A yellow wire is connected from the Y terminal on the thermostat to one of the connections on the contactor coil. The other side of the coil is connected to the C terminal on the transformer secondary.

System Operation: With the disconnect switch turned on with good fuses or circuit breakers there will be line voltage to the contactor **(C)** terminals L_1 and L_2. From these terminals line voltage is fed to the indoor fan motor **(IF)** and to the primary side of the transformer **(TR)**. The indoor fan motor will not operate because the indoor fan motor relay **(IFR)** contacts are in the open position.

There is 24V at the transformer secondary terminals and power is fed to the red terminal on the thermostat **(T)**. There will be a small amount of current flowing through the cooling anticipator and through the holding coil **(C)** of the contactor. The resistance of the cooling anticipator will prevent the contactor from pulling in, but a small amount of heat will be generated inside the thermostat. The purpose of this heat is to cause the thermostat to come on sooner than it would without the anticipator, to prevent system lag. The cooling anticipator is not adjustable.

When the system switch is in the on position and the thermostat demands cooling the, the cooling contactor coil **(C)** is energized and the contacts **(C)** close. The cooling anticipator is bypassed by the current because of its high resistance. The contactor coil **(C)** is energized which closes the contactor contacts **(C)**. When the contactor contacts close line voltage is fed onto both the run winding **(R)** and the common winding **(C)** of both the compressor motor and the condenser fan motor. The run capacitor **(RC)** connected between the run **(R)** and common **(C)** terminals of both motors adds an out-of-phase current to the motor auxiliary windings and provides the necessary starting torque for the motor to start. It remains in the circuit to help the operating efficiency of the motor.

When the cooling contacts in the thermostat **(T)** close 24V power is also directed to the indoor fan relay **(IFR)** through the fan contacts in the thermostat which are connected to the indoor fan relay coil by a green wire. The indoor fan relay contacts close to complete the line voltage circuit to the indoor fan motor **(IF)**. Should the operator want the indoor fan to operate continuously, the fan switch could be placed in the on position which would energize the indoor fan relay coil continuously and allow line voltage to the motor regardless of the compressor operation.

The system will remain in this mode until the thermostat is satisfied. Then it will de-energize the indoor fan relay coil **(IFR)** and the contactor coil **(C)**. The indoor fan, the condenser fan, and the compressor would all stop. The system would remain in this mode until the thermostat again demands cooling. All of the contacts return to their normal, de-energized, position.

Simple Heating and Cooling System Schematic: During the normal everyday service work of a technician both the heating and cooling systems and their combined wiring diagrams will be used to determine system operation and help in diagnosing troubles with the system. The technician must be familiar with reading them to be successful in this vocation. Figure 66-18 is a combination of the heating and cooling diagrams discussed previously. This is usually how they will be seen in the field.

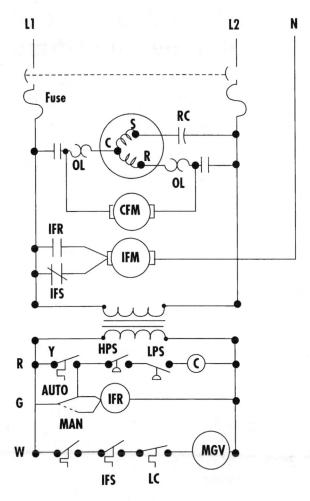

Figure 66-18 Simple heating and cooling unit schematic.

Unit 67: Pictorial Wiring Diagrams

A pictorial diagram is also sometimes referred to as a line diagram or a label diagram. Their purpose is to indicate the actual arrangement of the components in the control panel and the actual wiring connections made to the components. They are seldom drawn to scale and components that are not actually inside the control panel are placed outside the diagram and labeled. Pictorial diagrams are very useful in locating a specific component for testing or replacement. Especially when there are many relays used. The components are shown as they really appear, with the terminals in the proper place and the wiring color code to follow. See Figure 67-1.

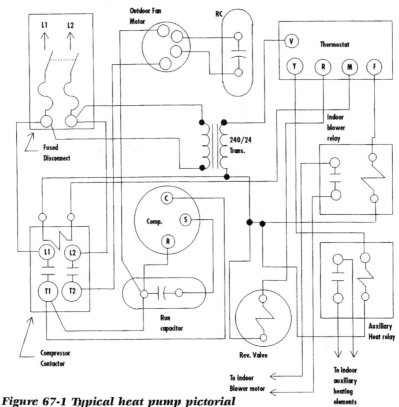

Figure 67-1 Typical heat pump pictorial diagram Basic wiring symbols.

When both a pictorial diagram and a schematic diagram are included on the same drawing it is usually referred to as a factual diagram. Some manufacturers provide factual diagrams for complicated systems and others provide them on all their units.

Unit 68: Installation Wiring Diagrams

Installation diagrams are not as detailed as the schematic of the pictorial diagram because the installation technician does not need to have this information in order to make the necessary installation connections. This type of diagram provides required information on color codes, fuse size, wire size, and breaker size, and terminals. See Figure 68-1.

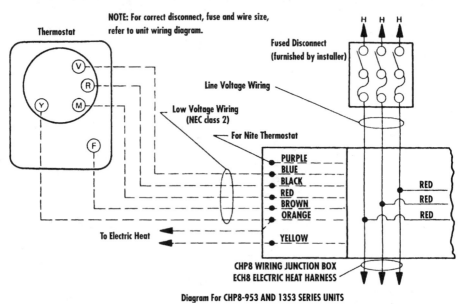

Diagram For CHP8-953 AND 1353 SERIES UNITS

Figure 68-1 Typical installation diagram for CHP8-953 and 1353 series units. (Courtesy of Lennox Industries.)

Notice that this type of diagram shows very little of the internal wiring or connections and is, therefore, almost useless to the service technician.

Unit 69: Gas Burner Controls

Gas burner controls are basically safety devices. They control the gas that flows into the combustion area of the furnace. Any time that gas is flowing into the combustion zone cannot be ignited safely these controls are designed to stop the gas flow. This is to help prevent an explosion should there be an accumulation of gas in the heat exchanger. In most common gas furnaces there are two ways that the gas is ignited as it leaves the main burner: by an automatic pilot and by high-voltage spark.

AUTOMATIC PILOT BURNER: The automatic pilot burner is used to ignite the gas as it leaves the main burner. The pilot gas may be ignited in two basic ways: by direct spark, and hot surface ignition.

Direct Spark (DSI): The direct spark method uses a high voltage to jump a gap between the ignitor-sensor and the pilot burner which is the ground for the ignition system.

When the thermostat demands heat, electricity is directed to the ignition module where the voltage is increased to around 10,000 volts. This high voltage then jumps the gap between the pilot burner/ignitor sensor rod and the pilot burner. This high-voltage spark is hot enough to ignite the gas escaping from the pilot burner. When the flame is sensed by the current flowing through the flame, a circuit will be completed to the ignition module. The main gas valves will then be energized to allow the main gas to enter the combustion zone. It must be noted that the pilot flame must continuously touch both the ignitor sensor rod and the pilot burner because this is part of the electric circuit. Any time the flame does not touch both surfaces, the circuit is interrupted and will cause a safety shutdown of the furnace. The pilot flame will then safely ignite the gas escaping from the main burner. If a flame is not sensed in a given period of time the module will shut down the unit. Some modules will allow a second trial for ignition and then shut the unit down which will require that a technician make the necessary repairs.

When the thermostat is satisfied, it will interrupt the control circuit to stop the main gas and the pilot gas flow. This intermittent burning pilot saves gas by not burning when it is not needed.

Hot Surface Ignition (HSI): Hot surface pilot ignitors are made by placing a resistor inside a silicone carbide casing. When the thermostat demands heat the ignition module causes the resistor inside the casing to become hot enough to light the gas escaping from the pilot burner. When the flame is sensed by either a sensing probe or by the hot surface ignition, the ignition is turned off and the main gas valve is energized to allow gas to flow into the combustion zone where it is ignited by the pilot flame. See Figure 69-1.

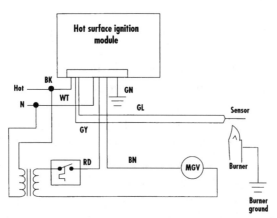

Figure 69-1 Typical hot surface ignition system.

The exact operating sequence will be determined by the equipment and the control manufacturers. Their data should be reviewed before making any repairs to the system.

MAIN BURNER IGNITION: There are generally two methods used to ignite the main burner gas: the pilot burner discussed above and the direct spark ignition method. Since we have already discussed the pilot lighting methods we will only discuss the direct main burner ignition method at this time.

Direct Ignition System: In an effort to conserve energy most manufacturers have started using the direct main burner ignition system. This method completely eliminates the use of pilot burner gas. These systems use either a high-voltage spark across the main burner ports or a hot surface as discussed above, to ignite the gas as it escapes from the burner.

Sequence of Operation: When the thermostat calls for heat, the ignition module makes a safe start check to test the internal module components for a flame simulating condition. Should there be a flame simulating failure detected, the ignition procedure will not start. If there is no flame simulated failure detected, the module begins a safety lock-out timing, powers the ignitor, and opens both valves in the gas control to allow gas to flow to the main burner.

Direct ignition systems may use either a spark ignitor or a hot surface ignitor to light the gas. When the spark ignition system is used, the ignitor starts to spark and the gas control valve opens at the same time. When the hot surface ignition system is used, the module allows about 30-45 seconds for the ignitor to warm up before opening the gas control valve and admit gas to the main burner.

Most ignition modules used on power burners have a prepurge cycle that must be completed before the ignition process is started. During the prepurge period, the combustion blower runs to remove all unburned fuel from the combustion zone. The prepurge period lasts for about 30-45 seconds. Most models use a separate air proving switch to make certain that the ignition module starts the timing period only after the proper air flow is established.

When the ignition period begins the burner must be lit and the flame must be proven within the safety lockout timing period of the module. If the flame is not proven, the ignitor stops, the gas valves close, and the system must be manually reset to restore operation. Some ignition modules will allow for several tries for ignition before the system is completely locked out.

Operation of the flame rectification method of flame sensing is as follows: When the main burner gas is ignited, electric current flows from the sensor through the flame to the burner head to ground. This is possible because of the carbon in the flame. Because of the difference between the size of the burner and the sensor, the electric current flows mainly in only one direction. Thus, making the electric current and pulsating direct, or rectified, current. This flow of current signals the ignition module that the flame has been established. In response to the current flowing through the flame, the ignition process is stopped and the main burner gas remains ignited.

There is a minimum amount of this dc current required to provide the signal to the ignition module and keep the gas control valves open. Should this current fall below the minimum value required, or becomes unsteady, the gas control valves will close, stopping the flow of gas to the main burner. The ignition module will perform another safety check to restart the ignition sequence.

CONTROL VALVES: Most manufacturers are now using redundant type gas control valves. This is to help in preventing furnace overheating because of a stuck open gas control. It would be a very rare occasion for two valves to stick open at the same time. Thus, the redundant, or dual, gas control is also a safety device for gas furnaces.

Operation: When the thermostat calls for heat, the ignition module energizes the pilot section of the 1st operator in the control valve. When the flame is proven the 2nd operator will open and supply gas to the main burner. Should the flame not be proven in the prescribed amount of time, the system will be locked out on safety.

When the thermostat is satisfied, both valve coils are de-energized at the same time to shut of the supply of gas immediately. There are different redundant valves used for standing pilot systems, intermittent pilot systems, and for direct ignition systems. If replacement is required be sure to use the correct one or the system could either not operate or operate in a dangerous condition.

Unit 70: Oil Burner Controls

The functions of any type of control system is to provide safe, automatic operation of the equipment. They are to function in response to changes in the space temperature and stop the unit when an unsafe condition occurs. The following is a description of the basic controls used for oil burner operation.

PRIMARY CONTROL: This control is generally considered to be the heart of the oil heating system. Its many functions are to control the burner motor, ignition transformer, and the oil valve when the thermostat demands heat. In most instances the high limit control is installed in the electric power line to the primary control so that when the limit control senses an unsafe condition the complete unit will be shut down. There is also a sensor that is a part of the primary control assembly. Its purpose is to monitor the burner flame during startup and through the complete burner on cycle. Should the flame fail for any reason, or if there is a power failure, the primary control will shut down the oil burner. The method used to sense the flame is determined by the type of primary controller used and its location in the heating system.

TYPES OF PRIMARY CONTROLS: There are basically three types of primary controllers used today. They are:

1. Burner-mounted with a cad cell (light sensitive) flame detector.

2. Burner-mounted with stack mounted (thermal) flame detector.

3. Stack-mounted with thermal sensor included (one piece).

The following terms define the action of the primary control:

Constant Ignition: The ignition comes on when the burner is energized and will stay on as long as the burner is firing.

Intermittent ignition: The ignition comes on when the

oil burner is energized. It is turned off after the main burner flame has been established or after a preset ignition timing period.

Nonrecycling Control: This control attempts to restart the burner immediately on the loss of the flame. The ignition attempt continues until the control locks out on safety.

Recycling Control: This type of control shuts down the oil burner immediately on the loss of flame, then attempts to restart the burner one time before locking out on safety.

Cad Cell Primary Controls: This type of primary control also includes a light sensitive cad cell mounted so that it can "see" the flame of the oil burner. The cad cell has a reduced resistance when light is present and is used to complete the flame detection circuit. This prevents the primary control from going into a safety lockout. The fast response of the cell to light rays eliminates the time lag that is present in bimetal type sensors. This makes this type of control very advantageous in large oil burner applications.

CAD primary controls may include and integral aquastat controller to combine the actions of burner operation with water temperature or circulator control in hydronic heating systems.

Cad Cell Flame Detectors: This type of flame detector consists of a photocell and a holder and cord assembly. The flame detector is placed inside the air tube of the primary burner where it can view the flame. It is wired to the primary control.

The photocell is made of a ceramic disc coated with cadmium sulfide overlaid with an electrically conductive grid. The electrodes are attached to the disc and transmit an electrical signal to the primary control. The cell is completely hermetically sealed with glass and metal to prevent any deterioration of the cell.

When the cell is located in a dark area it has a very high resistance to electric current flow. However, when light becomes visible the resistance begins to drop to a very low value and the current is allowed to pass through it.

The amount of light emitted by an oil flame is in the response range of the cell, but the typical gas flame is above this range and the cad cell will not respond to it. However, it is possible, though not typical, to apply cad cells in specific gas burners to view a portion of the flame pattern which is rich in fuel. This is a practical installation if the cell can be installed to always view this special flame condition.

The refractories used in oil burner combustion zones have a wave length of about 7000 or more angstroms. Most of the cad cell response range is lower than 7000 angstroms.

When an operating oil burner is properly adjusted the cad cell resistance will be in the range of 300 to 1000 ohms. To make certain that the burner will continue to operate reliably, the resistance of the cad cell should be above 1600 ohms when the burner is operating. When a resistance greater than 1600 ohms is detected the cad cell may need cleaning or adjusting, or the burner flame may need to be adjusted.

NORMAL OPERATION: During normal operation of the R8119 when the thermostat calls for heat, relay 1K pulls in to energize the ignition circuit and start the burner motor. Plus the ignition timer starts heating.

When the cad cell sees a normal flame, relay 2K operates, shunting the safety switch heater. The timer switch shuts off the ignition in approximately 70 seconds.

When the thermostat is satisfied and signals the unit to stop, relay 1K drops out, de-energizing the ignition circuit, the oil valve, and the burner motor. Relay 2K returns to the starting position as soon as the cad cell senses that there is no flame. The timer also prevents a burner restart for 2 to 3 minutes.

Unit 71: Basic Air Conditioning Control

INTRODUCTION: Modern air conditioning systems include control systems that range from the very simple to the very complex. Depending on the requirements of the system. Technically, air conditioning includes heating, cooling, humidification, dehumidification, air circulation, and air cleaning. In most residential systems the control circuits are very simple. However, in commercial/industrial type systems the control circuit can be very complex because of the special requirements for system operation. In most installations there are more safety devices included in the commercial systems because of the expense of the equipment replacement. In recent years, equipment manufacturers have tried to make residential systems less complex without omitting any of the safety features needed for safe operation. The advancements in control technology have made this trend possible. Solid state controls are the most popular type used in both residential and commercial equipment. However, solid state components are not all that make up the control circuit in air conditioning systems.

Each piece of equipment has its own control circuit which is different from any other system components. For example, the furnace will have a control circuit which is different from the one used in the condensing unit. The gas furnace will have a different type of control circuit from that used on an electric furnace. There will be some means of interlocking the heating and cooling sections of the systems together electrically so that they will operate in combination as desired.

The thermostat is the major operating control used in air conditioning systems. The equipment is protected by the safety devices that are included in the control circuit. Thus, the system operates automatically and safe without being manually controlled except when a problem arises, then the safety device may need to be reset and the system checked by a service technician. Or the on and off operation of the equipment may be operated manually by the system and fan switches located on the thermostat.

Most installation companies require that the installation technician know how to properly connect the control circuits together so that they will operate as desired. There are diagrams included in the installation packet that is packed with the equipment to guide the installation technician through the necessary steps. However, when the installation wiring diagram is not present, the technician must be able to properly connect the control circuit so that it will function properly.

It has been estimated that approximately 85% to 90% of the problems encountered with air conditioning and heating systems are electrical in nature. Therefore, it is necessary that all the personnel in this industry be familiar with the different types of control circuits used so that the customer can be properly instructed as to the correct use to get the most satisfaction out of the equipment.

BASIC CONDENSING UNIT CONTROLS: The condensing unit is that part of the system that is located out of doors on split systems. See Figure 71-1.

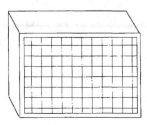

Figure 71-1 Basic air-cooled condensing unit.

The condensing unit consists of the compressor, condenser fan motor, the condenser, the contactor or starter, and any necessary safety devices such as the pressure controls and motor overload protectors. See Figure 71-2.

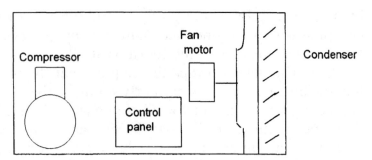

Figure 71-2 Air cooled condenser components.

The condensing unit is connected with refrigerant lines and the necessary power and control wiring to and indoor unit such as a heater, blower coil, etc. to make it operate as designed. Most modern condensing units are air cooled. Some are water cooled, but they are usually of the larger sizes. This is because of the extra expenses associated with a water cooled condenser and other problems that are inherent with the water cooled equipment.

The electric power wiring is connected to a fuse or circuit breaker panel on a dedicated circuit. The control circuit is usually connected to the indoor section of the system. Then both the condensing unit and the indoor section is properly connected to a temperature control device, such as a thermostat so that the system will operate automatically.

Simple Control Circuit for An Air Cooled Condensing Unit: In most installations, the condensing unit uses the transformer inside the indoor unit for the required 24 vac used in the control circuit. When a packaged (PTAC) unit is used the 24 vac is supplied by a transformer package located inside the unit cabinet. The indoor fan relay coil, and the compressor contactor coil are connected to the transformer and are controlled by the thermostat. The compressor contactor actually energizes or de-energizes both the compressor and the condenser fan motor when cooling is called for by the thermostat. The safety devices in either, or both, the compressor and fan motor may prevent that component from operating even though the contactor is energized, pulled-in. This control circuit contains the least amount of controls and is therefore the least expensive. See Figure 71-3.

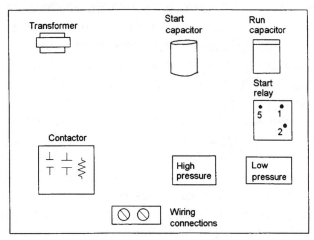

Figure 71-3 Simple control panel of an air cooled condensing unit.

Condensing units not equipped with a high-pressure control must be equipped with an internal relief valve. The purpose of this valve is to open and relieve the pressure should the discharge pressure reach an unsafe operating level.

The schematic wiring diagram for this type of control circuit is very simple. See Figure 71-4.

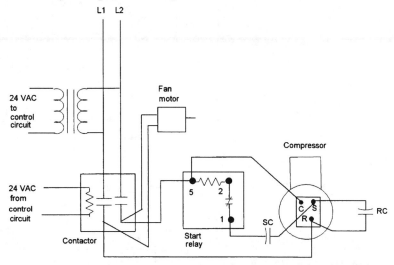

Figure 71-4 Simple wiring schematic for an air cooled condensing unit.

Operation: When power is supplied to the system through L_1 and L_2 the components are located between the two power wires. The transformer primary is energized and supplies 24 vac to the control circuit. When the thermostat demands cooling, the compressor contactor coil and the indoor fan relay coil are both energized. The indoor fan starts running and the compressor and condenser fan motor start running at the same time. As long as there is no safety problem, the system will operate in this manner until the thermostat is satisfied. When the thermostat is satisfied, its contacts open to de-energize the control system and allow the indoor fan relay coil and the contactor coil to return to their normal de-energized position. The system is now ready for another cycle when the thermostat demands.

Complex Control Circuit for An Air Cooled Condensing Unit: Manufacturers include on their more expensive units a more complex control circuit to protect the unit should an unsafe condition occur. The more complex control circuits include high-and low-pressure switches. See Figure 71-5.

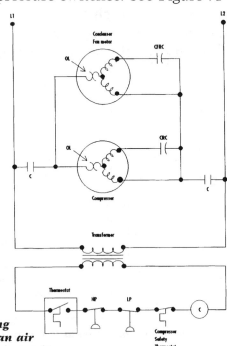

Figure 71-5 Complete wiring schematic for an air cooled condensing unit.

In this type of circuit, if any single control interrupts the current then the complete circuit will become inoperative and stop the condensing unit.

Some manufacturers also include an anti-cycling device to prevent the compressor from restarting before a predetermined period of time has lapsed. This is usually called an anti-short-cycling control. The purpose of this control is to prevent the compressor from restarting before the pressures have had time to somewhat equalize before restarting under an overloaded condition. This condition is known as short cycling. Some of these types of controls include a time clock that will automatically allow the unit to restart after a given period of time has lapsed. Others use a manually reset short cycling control that must be manually reset before operation can be resumed. This should be done by a qualified technician so that the cause of the problem may be found and repaired before damage can be done to the compressor.

Some of the larger commercial units make use of other type controls to provide the needed protection for their equipment. However, once the control circuit is understood these additional controls are not usually a problem for the technician in determining what controls are used in the circuit.

All equipment is shipped from the manufacturer with installation wiring diagrams included in the installation packet. In most cases it is usually easy to follow these diagrams and correctly wire the unit. See Figure 71-6.

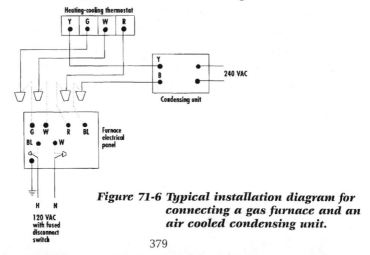

Figure 71-6 Typical installation diagram for connecting a gas furnace and an air cooled condensing unit.

TROUBLESHOOTING: When troubleshooting is required, the schematic diagram makes the process much simpler. When the schematic is used along with an understanding of the controls used in the circuit, there should be no problems in troubleshooting a residential air conditioning circuit. In most cases the commercial air conditioning schematic diagrams may require some study before a working understanding of the operating sequence is understood. Especially when the circuit is new to the technician.

FIELD WIRING REQUIREMENTS: All air conditioning systems require that some field power wiring be done after the system has been installed. Sometimes this wiring can be done by the installation technician, but usually it is done by a licensed electrician. However, the installation technician must have a working knowledge of this wiring so that he/she can make certain that it is done correctly.

All wiring that is done before the equipment leaves the factory is called factory wiring. In most units this includes all the wiring of the unit components to the field wiring terminals inside the control panel. The factory wiring is properly sized and usually includes some type of color code to aid in tracing the wiring should a problem occur. The required control wiring done in the field is usually done by the installation technician. The control wiring usually is available in a number of size 18 to 20 with several wires inside a single covering. The contained wires are color coded to aid in the wiring process and to aid in troubleshooting after the system is completely installed. Control wiring is available in the number of conductors from 2 to sometimes 10 or 12 conductors, depending on the number of functions required for proper operation of the system.

Unit 72:
Basic Refrigeration Controls

The control circuit used in refrigeration systems is usually line voltage for 240 volts ac. The basic refrigeration controls are the cold control, high-and low-pressure controls, condensing unit contactor, a defrost control, and a reversing valve or electric defrost control. Both the condensing unit and the indoor fixture, walk-in cooler, frozen food display or other types of storage cases are connected to the line (power) voltage for the system.

The display cases include the evaporator fan motors, the cold control, and the electric defrost coils if used. The condensing unit consists of the compressor, condenser fan motor, the pressure controls, and the reversing valve (if used). The defrost control is usually mounted on the wall close to the condensing unit, unless it is included in the unit proper.

OPERATION: The electric power is fed to the unit through lines L_1 and L_2. Each of the components are wired between these two lines.

When the cold control calls for cooling, the case fans, and the compressor contactor are energized at the same time. See Figure 72-1.

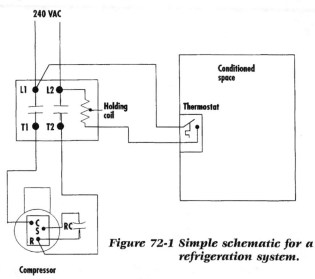

Figure 72-1 Simple schematic for a refrigeration system.

The case fans, the compressor, and the condenser fans are all started. The system will operate in this manner until the cold control is satisfied or the defrost control calls for a defrost cycle. When the cold control is satisfied the system will shut-down and wait for another cooling demand from the cold control.

When the defrost control demands a defrost cycle the case fans are stopped, the compressor and condenser fan are running, the reversing valve is energized to feed hot refrigerant to the evaporator coil rather than the condenser. The case fans are stopped to prevent blowing warm or hot air into the case and increasing its temperature unnecessarily. When all the frost is melted from the evaporator, the defrost control switches the reversing valve, the case fans are started, and the compressor and condenser fans continue running until the cold control is satisfied to stop operation of the unit.

The control circuit also contains devices such as high- and low-pressure controls, compressor overload protectors, fan motor overload protectors, running capacitors for the condenser fans and compressor, and starting capacitors for the compressor. Some of these are safety controls and others are used to improve the operation and efficiency of the unit.

Section 6: Using Schematic Diagrams In Troubleshooting

INTRODUCTION:

The different types of air conditioning, heating, and refrigeration require all types of control systems and circuits. Depending on the requirements of the system. The control systems that are used in the larger sizes of equipment must be more complicated because of the operation and the cost of repairs and equipment. In the smaller systems, from about five tons and less, the components and repairs are not as expensive. Also, the temperature control may be allowed to vary more than with the larger systems.

For economic reasons the larger systems usually include some type of capacity reduction on the compressor, which is not required on the smaller systems. Also the control systems used on the larger systems require that all of the components are operating properly or the system will be shut down for safety reasons and cost of the equipment and repairs.

Unit 73: Reading Advanced Schematic Diagrams (Single-Phase)

As with just about everything in this industry it would be impossible to introduce all the systems used. Therefore, we will discuss the general requirements of a control system by representing only a few in detail for reference. The specific equipment manufacturer's diagrams must be checked to make certain that the correct wiring diagram is being used when making repairs.

SINGLE-PHASE COOLING SYSTEM: First we will discuss the operating G350 sequence of the Lennox HS29 Single-Phase split system cooling unit: Refer to Figure 73-1.

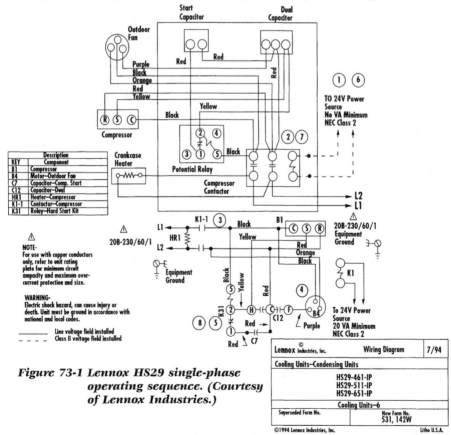

KEY	Description Component
B1	Compressor
B4	Motor—Outdoor Fan
C7	Capacitor—Comp. Start
C12	Capacitor—Dual
HR1	Heater—Compressor
K1-1	Contactor—Compressor
K31	Relay—Hard Start Kit

NOTE-
For use with copper conductors only, refer to unit rating plate for minimum circuit ampacity and maximum over-current protection and size.

WARNING-
Electric shock hazard, can cause injury or death. Unit must be ground in accordance with national and local codes.

——— Line voltage field installed
- - - - Class II voltage field installed

Lennox © Industries, Inc.	Wiring Diagram	7/94
Cooling Units—Condensing Units		
	HS29-461-IP HS29-511-IP HS29-651-IP	
Cooling Units—6		
Superseded Form No.	New Form No. 531, 142W	
©1994 Lennox Industries, Inc.		Litho U.S.A.

Figure 73-1 Lennox HS29 single-phase operating sequence. (Courtesy of Lennox Industries.)

A-HS29 "P" Voltage operation for HS 29-461/511/651 "P" voltage units. The HS29-211/261/311/411 "P" voltage units are not equipped with the hard start kit. The sequence is outlined by numbered steps which correspond to circled numbers on the proper diagram.

Note: The thermostat used may be electromechanical or electronic.

Note: The transformer in the indoor unit supplies 24 VAC power to the thermostat and the outdoor unit controls.

COOLING OPERATION:

1. When cooling is required the Y1 terminal in the thermostat receives power from the transformer.

2. The 24 VAC is supplied from the transformer in the indoor unit which passes through the thermostat first stage cooling contacts to energize the cooling contactor K1.

3. K1-1 N.O. contacts close, energizing terminal "C" of the compressor **(B1)** and the outdoor fan motor **(B4).**

4. The outdoor fan motor **(B4)** begins immediate operation.

5. The compressor **(B1)** is started. The hard start contactor K3-1 remains closed during the start-up period and the start capacitor C7 remains in the starting circuit. As the compressor gains speed, K3-1 is energized. When K3-1 is energized, the contacts open and the start capacitor C7 is taken out of the starting circuit.

END OF COOLING DEMAND:

6. When the cooling demand is satisfied, terminal Y1 is de-energized.

7. The compressor contactor K1 is de-energized.

8. Contact K1-1 opens and the compressor **(B1)** and the outdoor fan motor **(B4)** are de-energized and stop immediately.

Unit 74: Reading Advanced Diagrams (Three-Phase)

We will discuss the Lennox Three-Phase cooling system operating sequence in this unit. While three-phase units are not commonly used in residential and small commercial installations they are very popular in the larger systems. Refer to Figure 74-1.

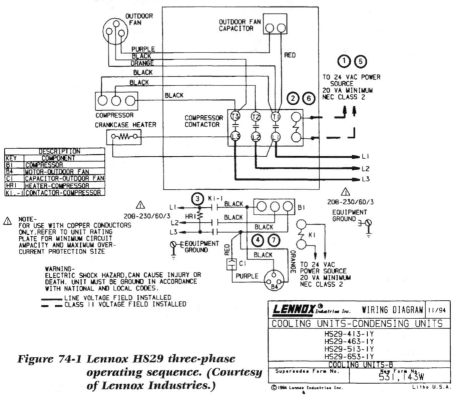

Figure 74-1 Lennox HS29 three-phase operating sequence. (Courtesy of Lennox Industries.)

HS29 "Y", "G" and "J" Voltage Operation Sequence:

This is the sequence of operation for HS29 "Y" voltage units. The HS29 "G" voltage sequence is the same; however, the "J" voltage units have an outdoor fan transformer. The sequence is outlined by numbered steps which correspond to circled numbers on the diagram.

NOTE: The thermostat used may be electromechanical or electronic. **NOTE:** The transformer in the indoor unit supplies 24 VAC to the thermostat and the outdoor unit controls.

COOLING:

1. When cooling is demanded, the Y1 terminal in the thermostat is energized.

2. 24 VAC then energizes the compressor contactor K1.

3. K1-1 N.O. contacts close to energize the compressor **(B1)** and the outdoor fan motor **(B4)**.

4. The compressor **(B1)** and the outdoor fan motor begin immediate operation.

END OF COOLING DEMAND:

5. When the cooling demand is satisfied, thermostat terminal Y1 is de-energized.

6. The compressor contactor coil K1 is de-energized.

7. Contact K1-1 opens and the compressor **(B1)** and the outdoor fan motor **(B4)** are de-energized and stop immediately.

Unit 75: Single-Phase Heat Pump Systems

Heat pump systems are somewhat more complicated than simple cooling systems because of the defrost cycle required to maintain unit efficiency. We will discuss the single-phase systems in this unit.

This is the sequence of operation for the HP29 "P" voltage units. The sequence is outlined by numbered steps which correspond to circled numbers on the diagram.

NOTE: The thermostat used may be electromechanical or electronic.

NOTE: The transformer in the indoor section provides 24 VAC power to the thermostat and the outdoor unit controls.

NOTE: *When the thermostat is switched to the cooling mode, 24 VAC is routed to the reversing valve coil. The reversing valve remains energized until the thermostat is switched to the heating mode. While in the heating mode, the reversing valve remains de-energized unless there is a call for a defrost cycle. In the defrost cycle, the reversing valve is controlled by CMC1.*

COOLING (Reversing Valve Energized):

1. The demand for cooling energizes the Y1 terminal in the thermostat.

2. 24 VAC then energizes the compressor contactor K1.

3. K1-1 N.O. contacts close, energizing the compressor **(B1)** and the outdoor fan motor **(B4)**.

4. The outdoor fan motor begins immediate operation.

5. The compressor **(B1)** begins operating. The hard start contact K3-1 remains closed during the starting period and the starting capacitor **(C1)** remains in the starting circuit. As the compressor gains speed, K3-1 is energized. When K3-1 is energized, the contacts open and the starting capacitor is taken out of the starting circuit.

END OF COOLING DEMAND:

6. When the cooling demand is satisfied, terminal Y1 is de-energized.

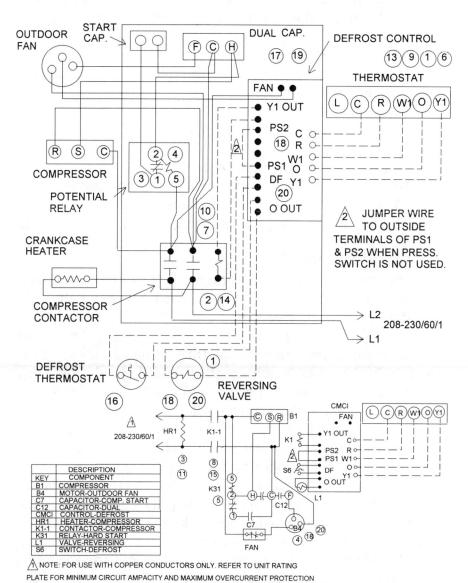

Figure 75-1 Lennox HP29 single-phase operating sequence. (Courtesy of Lennox Industries.)

7. The compressor contactor K1 is de-energized.

8. K1-1 contacts open to de-energize both the compressor **(B1)** and the outdoor fan motor. Their operation is stopped immediately.

FIRST STAGE HEAT (Reversing Valve De-energized):

9. The demand for heat is initiated at terminal W1 in the thermostat.

10. 24 VAC then energizes the compressor contactor K1.

11. K1-1 N.O. contacts close, energizing the compressor and the outdoor fan motor.

12. See Steps 4 and 5 above.

END OF FIRST STAGE HEAT:

13. When the heating demand is satisfied, terminal W1 in the thermostat is de-energized.

14. The compressor contactor K1 is de-energized.

15. K1-1 contacts open and the compressor **(B1)** and the outdoor fan motor **(B4)** are de-energized and stop immediately.

DEFROST MODE:

16. During heating operation when the outdoor coil temperature drops below $35°F \pm 4°F$, The defrost switch (thermostat) S6 contacts close.

17. The defrost control CMC1 begins timing the defrost cycle. If the defrost thermostat **(S6)** remains closed at the end of the 30, 60 or 90 minute time period, the defrost relay energizes and the defrost cycle begins.

18. During the defrost cycle CMC1 energizes the reversing valve and W1 on the terminal strip (operating indoor unit on the first stage heat mode), while at the same time de-energizing the outdoor fan motor **(B4)**.

19. The defrost cycle continues for 14 ± 1 minutes or until the defrost thermostat switch **(S6)** opens. When the defrost thermostat contacts open, the defrost control CMC1 loses power and automatically resets.

20. When the defrost control CMC1 resets, the reversing valve and W1 on the terminal strip are de-energized, and the outdoor fan motor **(B4)** is energized.

Unit 76: Three-Phase Heat Pump Operation

Three-phase heat pump systems are used mostly on small and large commercial installations. Their operation is somewhat different than the single-phase systems. The following is a description of their operating sequence. Refer to Figure 76-1.

HP29 THREE-PHASE OPERATING SEQUENCE

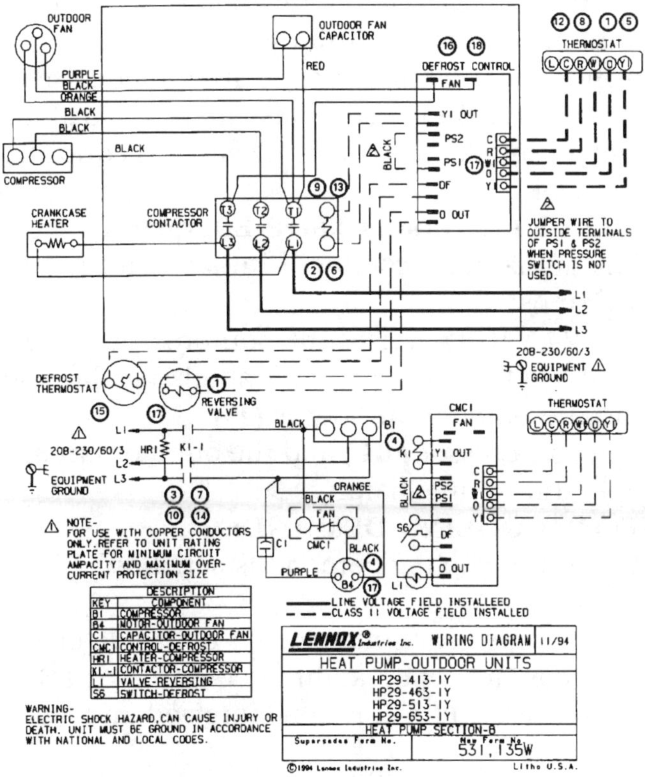

Figure 76-1 Lennox HP29 Three-phase operating sequence. (Lennox Industries.)

HP29 "Y", "G", AND "J" voltage OPERATION SEQUENCE

This is the sequence of operation for HP29 "Y" voltage. The HP 29 "G" and "J" voltage units are similar, but have a few additions. The "G" voltage units have an outdoor fan relay, while the "J" voltage units have the outdoor relay plus and outdoor transformer. The "Y" voltage unit sequence is outlined by numbered steps which correspond to circled numbers on the unit diagram.

NOTE: The thermostat used may be electromechanical or electronic.

NOTE: The transformer in the indoor section supplies 24 VAC power to the thermostat and the outdoor unit controls.

NOTE: When the thermostat is switched to the cooling mode, 24 VAC is routed to the reversing valve. The reversing valve remains energized until the thermostat is switched to the heating mode. While in the heating mode, the reversing valve remains de-energized unless there is a call for defrost. In the defrost mode, the reversing valve is controlled by the CMC1 control.

COOLING (Reversing Valve Energized):

1. The demand for cooling is initiated at the Y1 terminal in the thermostat.

2. The 24 VAC energizes the compressor contactor K1.

3. K1-1 N.O. contacts close energizing the compressor **(B1)** and the outdoor fan motor **(B4)**.

4. The compressor **(B1)** and the outdoor fan motor **(B4)** begin operation immediately.

END OF COOLING DEMAND:

5. When the cooling demand is satisfied, terminal Y1 is de-energized.

6. The compressor contactor K1 coil is de-energized

7. K1-1 contacts open and the compressor **(B1)** and the outdoor fan motor **(B4)** are de-energized and stop immediately.

FIRST STAGE HEAT (Reversing Valve De-energized):

8. The demand for heating is initiated at the W1 terminal in the thermostat.

9. 24 VAC energizes the contactor **(K1)** coil.

10. K1-1 N.O. contacts close and energize both the compressor and the outdoor fan motor.

11. See steps 4 and 5.

END OF FIRST STAGE HEAT:

12. When the heating demand is satisfied, terminal W1 in the thermostat is de-energized.

13. The compressor contactor coil K1 is de-energized.

14. K1-1 contacts open and the compressor **(B1)** and the outdoor fan motor **(B4)** are de-energized and stop immediately.

DEFROST MODE:

15. During the heating operation when the outdoor coil temperature drops below $35°F \pm 4°F$ the defrost switch (thermostat), S6 contacts close.

16. The defrost control CMC1 begins timing. If the defrost thermostat **(S6)** remains closed at the end of the 30, 60, or 90 minute time period, the defrost relay energizes and the defrost cycle begins.

17. During the defrost cycle the CMC1 control energizes the reversing valve and W1 on the terminal strip (operating the indoor unit on the first stage heat mode), while de-energizing the outdoor fan motor **(B4)**.

18. The defrost period continues for 14 ± 1 minutes or until the defrost thermostat switch **(S6)** opens. When the defrost thermostat opens, the defrost control CMC1 loses power and automatically resets.

19. When the CMC1 control resets, the reversing valve and W1 on the terminal strip are de-energized. The outdoor fan motor **(B4)** is energized.

Unit 77: Self-Contained Air Conditioning Units (PTAC)

The control systems used on PTAC units are fairly simple in the smaller sizes and become more complex as the size of the system increases. The larger units require a more complex control system because of zoning requirements that are sometimes desired or needed. Usually some type of interlock is used on these control systems to prevent the compressor from operating when the indoor or outdoor fan motor or the water pump, on water cooled systems, is not operating properly.

PTAC units can be used in combination with either gas or electric heating. Usually the heating equipment must be installed in the field, or they may be ordered from the factory with it already installed. In some cases it may be more feasible to use a hot water coil when sufficient hot water or steam is available for use in comfort heating.

The wiring diagram of a small air cooled PTAC system is fairly easy to read and understand. See Figure 77-1.

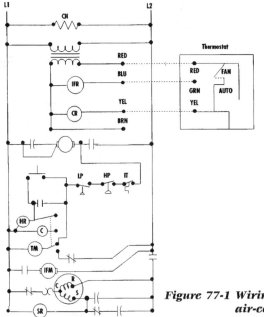

Figure 77-1 Wiring schematic of a small air-cooled PTAC unit.

Notice that this particular diagram has a legend indicating the different types of controls used in the circuit. A contactor is used to control the operation of the compressor, and the condenser fan motor. The evaporator fan motor is controlled by a fan relay which is energized by the low voltage circuit. All the required safety controls are also included in the control circuit to properly protect the system components.

The unit on which the diagram shown in Figure 77-1 would usually be used in a residence or a small commercial installation. The thermostat shown in this schematic is located at the bottom of the diagram and is what controls the operation of the system. The fan switch on the thermostat may be placed in either the "FAN" position or the "AUTO" position. In the "FAN" position it will operate continuously whether cooling or heating is required. When placed in the "AUTO" position it will run only when heating or cooling is required. The indoor fan is energized through the fan relay contacts when the relay coil is energized from the thermostat.

The control voltage passes through several control system components to energize the compressor and cause it to operate. The first step is for the thermostat to demand cooling, causing the control relay contacts to close. When these contacts close the outdoor fan motor and the contactor coil are energized through the holding relay and the required timer circuits. When the contactor contacts are closed, the compressor will begin operation. Should any one of the safety controls not be in the proper position the compressor will not operate until the trouble is located and corrected.

Section 7: Ice Makers

INTRODUCTION:

There are two basic types of ice makers, domestic and commercial. There are several differences between the two. In domestic models, a tray or mold is filled with water and allowed to freeze. The different makes use different ways to determine when the water is frozen and to remove the ice from the tray or mold.

Commercial type ice makers recirculate the water during the freezing cycle. The purpose of recirculating the water is because pure water freezes faster than unpure water which contains various types of minerals. The ice made in commercial ice makers is usually more pure and more clear than that made in domestic ice makers. This refers to cuber type ice machines and not flaker type units. Flaker type units scrape the ice from the evaporator with an auger as it is frozen.

Commercial ice machines can also be divided into two different types. They may be either of the cube type or the slab type. However, both types are usually called cubers. The cube type allows the ice cubes to drop directly into the storage bin during the harvest cycle. The slab types use a wire grid which is heated by electricity to cut the slab into squares or rectangle cubes.

Unit 78: Domestic Ice Makers

Domestic ice makers are located in the freezer compartment of the refrigerator-freezer. The ice maker is filled with water from the water supply piping. When the water contains many minerals, a filter is sometimes placed in the make-up water line to the ice maker. A solenoid valve is also installed in the make-up water line to the ice maker. The solenoid valve is opened by the electrical circuit to fill the ice tray. The tray is given a time period to be filled. See Figure 78-1.

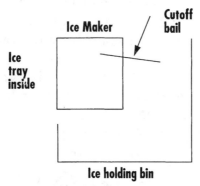

Figure 78-1 Location of ice holding bin.

When the temperature of the freezer compartment reaches the desired temperature the water is frozen into ice cubes. When a predetermined amount of time has passed, the ice maker enters what is known as the harvest cycle. Thus, the ice is removed from the ice maker and deposited into the ice bin located below the ice maker. Some makes use a wire bail fastened to a switch located in the ice maker so that when the ice bin is full the ice maker will automatically stop making ice. Each time there is a harvest cycle the bail is raised to allow the ice to drop into the holding tray. When the ice maker goes into another fill cycle the bail is allowed to fall to the top of the ice in the bin. When the bin is full the bail cannot drop low enough to allow the switch to close and start another freeze cycle. It will remain in this position until some of the ice has been removed, settled, or melted to allow the bail to fall further towards the switch closing position.

Another method of stopping ice production when the bin is full is to have the bin setting on a switch so that when the weight has reached a predetermined value a switch will be opened to stop ice production. When some of the ice has been removed or has melted to a predetermined weight, the switch will close and ice production will resume.

There are two general methods of loosening the ice from the ice maker. One is to twist the tray much like when we twist an ice tray to remove the ice from it when an ice maker is not used. The second method is to place an electric heater inside the ice mold to loosen the ice by heating the mold so that it can be mechanically removed from the mold.

TWISTING TRAY TYPE: When this type is used a small motor is used to twist the ice tray to loosen the ice cubes. At the same time the tray is turned about 90° so that the ice will fall into the ice bin. See Figure 78-2.

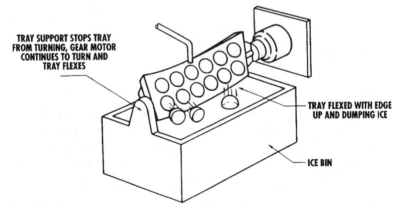

TRAY SUPPORT STOPS TRAY FROM TURNING, GEAR MOTOR CONTINUES TO TURN AND TRAY FLEXES

TRAY FLEXED WITH EDGE UP AND DUMPING ICE

ICE BIN

Figure 78-2 Twisting motion of ice tray.

When the ice has been removed from the freezing tray the motor will release and allow the tray to return to its horizontal position. The tray is refilled with water and the cycle is restarted.

HEATED MOLD TYPE: This type of ice maker allows

the ice to freeze in the mold and then electricity is applied to a heater located in the freezing mold. When the mold reaches a predetermined temperature the resistance heater is energized to soften the surface of the ice next to the mold so that it can be mechanically removed.

The water enters this type of ice maker just the same as with the flexible tray type. When the mold is full of water the freeze cycle is started and the mold heater is de-energized. This is the freezing cycle. See Figure 78-3.

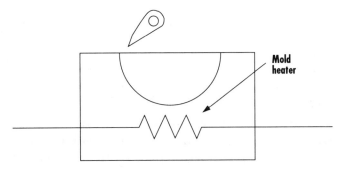

Figure 78-3 Freezing cycle.

When the ice mold reaches a predetermined temperature the heater is energized to soften the surface of the ice next to the mold and a small electric motor is started which causes small plastic fingers to apply pressure on the top of each ice cube. See Figure 78-4.

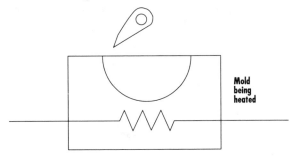

Figure 78-4 Defrosting cycle.

When the ice is released from the mold the plastic fingers force the ice out of the mold cavity by the motor and it falls into the storage bin. See Figure 78-5.

Mold heater

Figure 78-5 Harvest cycle.

The ice molds are again filled with water and the freeze cycle is restarted.

This type uses a bail in the ice storage bin. When the harvest cycle is in progress the bail is raised out of the way of the falling ice cubes. When the harvest motor has completed its cycle the bail is allowed to return into the storage bin. When the bin is full, the bail will not fall low enough to allow the bin switch to close to restart the freeze cycle again. The ice maker will remain off until some of the ice is removed, melted, or settled in the storage bin. In which case the freeze cycle will be restarted.

Some models use electronic circuits to automatically freeze and harvest the ice in domestic ice makers. The controls may look different from the electric models but the sequence of operation is the same.

Unit 79: Commercial Ice Makers

The following discussion is of the Ice-O-Matic model C-61 A&W (Air and Water cooled). This is a cuber type ice maker electrical sequence of operation. The flaker type commercial ice machines will have a little different type of wiring diagram but when you have mastered the wiring diagram of this type the flaker type will be just as easily read. In commercial ice machines water is constantly flowing over the freezer plate or the cube mold until sufficient ice is frozen to call for a harvest cycle.

FREEZE CYCLE 1ST STEP: Refer to Figure 79-1.

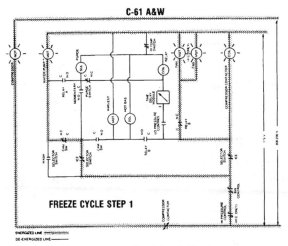

Figure 79-1 Freeze cycle 1st step electrical diagram.
(Courtesy of Mile High Equipment Company.)

When the ice storage bin control is closed, and the selector switch is in the ice position, electric power is supplied to the contactor coil, energizing the contactor and supplying power to the compressor motor. The contactor coil will remain energized and the compressor remains running until the bin control opens or the selector switch is turned to off or wash.

The water pump is supplied power through the selector switch and the normally closed contacts of the cam switch.

The fan motors are supplied power through the normally closed contacts of the relay.

FREEZE CYCLE 2ND STEP: Refer to Figure 79-2.

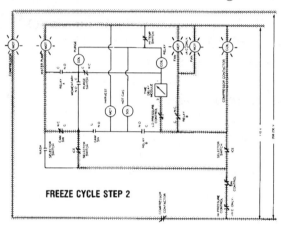

Figure 79-2 Freeze cycle 2nd step electrical diagram. (Courtesy of Mile High Equipment Company.)

At this time enough ice has formed on the evaporator to lower the refrigerant suction pressure sufficiently to close the low pressure control (the control closes at 30 p.s.i. on the C-61) energizing the Time Delay Module which starts its time delay mechanism. All the other components that were energized in step 1 are also energized in step 2.

DEFROST CYCLE 3RD STEP: Refer to Figure 79-3.

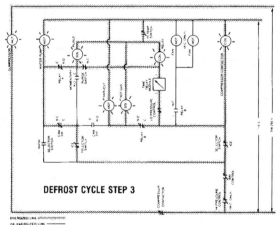

Figure 79-3 Defrost cycle 3rd step electrical diagram. (Courtesy of Mile High Equipment Company.)

The time mechanism in the time delay module is now finished timing and supplies power to the relay coil. By energizing the relay coil power is supplied to the hot gas valve, harvest motor, and the water purge valve solenoid, and de-energizes the fan motors.

Note that the rise in the suction pressure during the defrost period has opened the contacts in the low pressure control. The relay will remain energized as long as the cam switch remains in the normally closed position.

HARVEST CYCLE 4TH STEP: Refer to Figure 79-4.

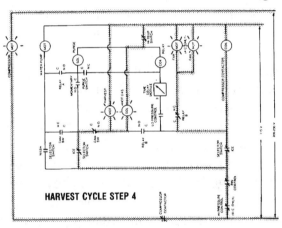

Figure 79-4 Harvest cycle 4th step electrical diagram. (Courtesy of Mile High Equipment Company.)

The probe assist motor has now moved forward enough to make the normally open contacts of the cam switch. Power to the relay coil is now interrupted and the relay has reset. The fan motors are now energized. The harvest motor and hot gas valve will remain energized until the cam switch returns to a normally closed position. At that time the harvest motor and hot gas valve will be de-energized, and the water pump will be energized. The machine will then start another freeze cycle.

Section 8: Solid-State Electronics

INTRODUCTION:

Solid-state electronics has gained acceptance by almost all equipment manufacturers since entering the air conditioning, heating, and refrigeration industry. Most manufacturers readily accepted their concept because of the reliability they provide. They are used as temperature sensors, pressure sensors, motor speed control has become more accurate and reliable. Solid-state controls are used for motor protection, defrost control for refrigeration and heat pump systems. they are more accurate and react more quickly than the electro-mechanical type controls used on the earlier units, and they are smaller in size.

Solid-state controls are popular in multi-zone control applications in large buildings because they have completely eliminated the erratic control of the individual zones. In many of this type of installation motor speed control is used to vary the volume of air delivered to the conditioned space to aid in controlling the temperature.

Solid-state controls used in large commercial refrigeration control systems provide better control to the complete system. Solid-state defrost systems are very popular in these types of installations because they provide more accuracy and reliable control of the defrost cycle. Some use variable speed control to aid in temperature control and system economy.

In the smaller type units used for refrigeration and air conditioning and heating use solid-state controls for a specific purpose. That is each control has a definite purpose. Not all the manufacturers of these types of systems have completely changed to solid-state control systems. However, they are rapidly approaching total solid-state control

Systems. The newer energy requirements have been achieved through the use of solid-state control systems. In addition to energy savings they are used to protect motors, sense refrigerant pressures inside the system, and to sense temperatures in certain parts of the system as well as in the conditioned space. Heat pump systems are now using the solid-state defrost control type system almost entirely. Furnaces are controlled better through the use of solid-state control systems.

To many technicians the term "solid-state" causes some apprehension. However, they need to be serviced just as the electro-mechanical type systems do. To overcome this apprehension, the technician has only to consider them as being merely a group of circuits used to control a particular system just as the electro-mechanical control systems do.

The technician who wishes to become proficient in this industry must learn to service the solid-state modules that are being used on air conditioning, heating, and refrigeration systems. In most instances the service will include the complete module rather than an individual circuit because the module incorporates several circuits and provides several different functions for the system. Generally it is the complete module that is replaced rather than some individual component of the module. When it is determined that a module is defective it is replaced and sent to the factory for repairs. Most service shops are not equipped to repair these modules, therefore it is much cheaper to send them to a repair facility rather than attempt the repair in-house. In any case the technician must learn the basics of solid-state and how it is applied to the control systems used in the equipment so that troubleshooting will be easier and more accurate. However, it is not necessary for the technician to become and electronics technician in order to effectively service solid-state control systems.

Unit 80: Semi-Conductor Materials

A semiconductor material is one that has only four valence electrons. Good conductors have less that half of their valence electrons and insulators have more than half of their valence electrons. A semiconductor will conduct electricity better than an insulator but not as well as a conductor. The most popular materials used as semiconductors are germanium, silicon, and selenium. When some of these materials are combined they share electrons, so that their valence shells are more full than before they were combined. From this we can see that pure semiconductor materials can be good insulators; however, they are good insulators only at absolute zero temperature. At higher temperatures, heat causes some of the valence electrons to become free and the material will then become a semiconductor.

Silicon is the most popular semiconductor material used in these types of control devices. When pure silicon is mixed with some type of impurity it becomes either a P-type silicon or an N-type silicon, depending on the type of impurity added to it. The P is an electrically positive and the N is an electrically negative material. When a P-type impurity material is added, the crystalline structure is modified, leaving a movable hole for an electron in the atom. See Figure 80-1.

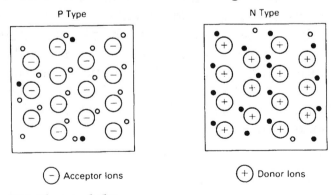

Figure 80-1 Electron holes.

When an N-type impurity is added, the structure is modified so that and extra movable electron is available. See Figure 80-2.

N Type

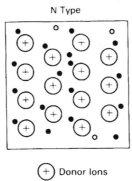

$(+)$ Donor Ions

Figure 80-2 Extra movable electrons available.

The separate P- and N-type materials alone have very little use in HVAC&R control circuits. However, when a junction is formed of both the P-type and the N-type material in such a manner that the crystalline structure is not broken, a P-N junction is formed. See figure 80-3.

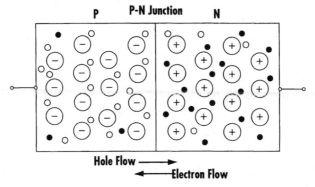

Figure 80-3 P-N junction.

Many of the semiconductor devices are made up of at least one P-N junction and it is considered to be the basic building block of semiconductor devices. The P-N junction has the unique ability to easily conduct current in only one direction.

Unit 81: The P-N Type Junctions

Before we tackle a study of a P-N junction, we must first consider the condition of the separate P-type and N-type materials before they are joined. See Figure 81-1.

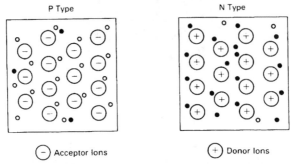

$\ominus$ Acceptor Ions $\oplus$ Donor Ions

Figure 81-1 P-type silicon before joining.

The two pieces cannot be simply pushed together. The crystalline structure must remain together at the point of the junction. The P-type material is made up mostly of silicon atoms, a small number of acceptor atoms, holes (or majority carriers), and a few free electrons. The acceptor atoms are shown as negative ions because at room temperature all of them will have accepted a valence electron from a neighboring atom to fill its empty covalent bond. These acceptor atoms are fixed in the crystal structure of the silicon and cannot move. The number of majority carriers depends on the number of acceptor atoms. The number of electrons (minority carriers) is dependent on the material temperature.

The N-type material is made up mostly of silicon atoms, a small number of impurity atoms, free electrons (or majority carriers), and a small number of holes, because of the thermal generation of electron-hole pairs. See Figure 81-2.

N Type

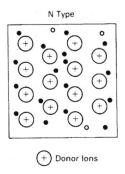

(+) Donor Ions

Figure 81-2 N-type silicon before joining.

The silicon atoms are not illustrated in the figure and should be thought of as a continuous crystal structure. The positive ions represent donor impurity atoms because at room temperature all have released one electron. These donor ions are fixed in the crystalline structure and cannot move. The number of majority atoms is dependent on the number of donor atoms. The number of holes (the minor carriers) is dependent on the material temperature.

When the P and N areas are joined together to form a single crystal structure, there is a high concentration of holes (the majority carrier) in the P region. See Figure 81-3.

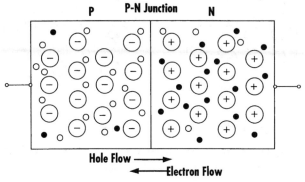

Figure 81-3 P-N crystal formed by joining the P and N regions.

There is a very low concentration of holes in the N region. The holes are the minority carriers in that region. It is the difference in concentration of holes in the semiconductor material that allows the flow of electric current. The holes in the P region will immediately begin diffusing from the area of high hole concentration across the junction into the region having a low concentration of holes. Also, there is a high concentration of free electrons, the majority carrier, in the N region. In the P region, there is a low concentration of free electrons, the minority carrier. Thus, the electrons are diffused from the N region to the P region across the junction.

As the P region holes diffuse across the junction, they encounter many free electrons. Each diffusing hole combines immediately with a free electron and then disappears. The electrons that are diffusing from the N region across the junction to the P region immediately combine with the available holes and disappear. Not all of the holes from the P region or all of the free electrons from the N region diffuse across the junction. It is only those majority carriers near the junction that diffuse across. This is possible because there is a spaced-charge region across the junction of the P-N crystal after some of the majority carriers have diffused across the junction and combined. See Figure 81-4.

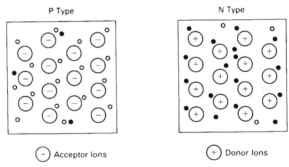

P Type   N Type

$\ominus$ Acceptor Ions $\oplus$ Donor Ions

Figure 81-4 P-type and N-type silicon before joining.

Only the fixed ions remain close to the crystal junction with negative ions on the P side and positive ions on the N side. There are no charge carriers in this region. It is these fixed charges close to the junction, that are caused by the diffusion of major carriers, which have a discouraging effect and eventually prevent any further diffusion. Should a hole from the P side attempt to diffuse across the junction, it is repelled by the positively charged donor ions which are on the N side of the junction. Also, any electrons trying to move across the junction from the N side are repelled by the negatively charged acceptor ions on the P side of the junction.

The space-charge region is a repelling electrical force. It is fixed charges on the opposite sides of the junction that produce a potential (electrical) barrier, the same that a battery would produce. If we measured the voltage across the space-charge region we would find a potential equal to this barrier. See Figure 81-6. It would be approximately 0.3 V for germanium and 0.7 for silicon.

On either side of the junction away from the space-charge region, the charges are concentrated exactly as they were before joining of the P-N regions. The charges, both positive and negative, are equally distributed. Thus, the net density of the charge is almost zero in all areas except in the space-charge region. See Figure 81-5.

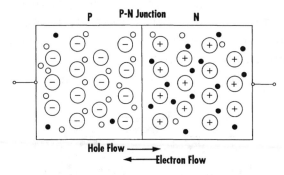

Figure 81-5 P-N crystal formed by joining the P and N regions.

In the space-charge region there exists a large positive charge on the N side of the junction and a large negative charge on the P side of the junction.

The P-N junction, because of the space-charge region, offers a resistance to any carriers attempting to move across the junction. Thus, for a free electron in the N region to cross the junction to the P region, it must gain enough energy to overcome this barrier. Also, for a hole in the P region to cross over into the N region a valence electron in the N region must overcome the barrier to fill this hole and have the hole appear on the N side. Because of the heat energy gained at room temperature, some of the majority carriers will overcome the barrier and move across the junction. This small amount of movement results in a majority-carrier movement across the P-N junction. This movement of current is the sum of the hole current plus the electron current. In contrast, this movement of current is exactly balanced by a current of equal value set up in the opposite direction and formed by the minority-carrier current. It is the attraction of the space-charge region that causes the attraction of the space-charge region that causes this minority-carrier current. See Figure 81-6. Should an electron-hole pair be produced by the heat energy in the space-charge region, the electron will be attracted toward the N region by the positively charged ions and the hole will be attracted toward the P region by the negatively charged ions. This attraction causes a minority-carrier drift across the P-N junction in a direction opposite to the flow of the majority-carrier defusion current. When there are no external power connections on the P-N crystal, the majority-carrier diffusion current and the minority-diffusion current cancel each other out, resulting in a net junction current of zero. See Figure 81-6.

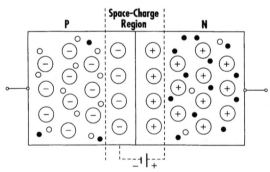

Figure 81-6 P-N crystal after some of the majority carriers have diffused.

When an external power source is applied to a P-N junction, it is said to be biased and the applied voltage is said to be a bias voltage. When no external power source is applied to the P-N junction, the junction is unbiased.

When working on solid-state circuits the technician must

be aware that the circuit operating mode is given explained using the convention current flow theory which is opposite to the electron flow theory. *In conventional current flow the theory assumes that the current flows from the positive to the negative of the power source. The electron current flow theory assumes that the current flow is from negative to positive.* The major reason for this theory, such as; the ground side of a circuit is generally considered to be zero volts in an electronic circuit. Any voltage that is greater than the ground voltage is considered to be a positive voltage. Also, all of the arrows indicated on electronic diagrams are pointed in the direction of conventional current flow. A diode is forward biased only when a positive voltage is applied to the anode and a negative voltage is applied to the cathode. See Figure 81-7.

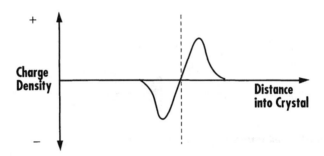

Figure 81-7 Charge profile of a P-N crystal.

Using the conventional current flow theory, the current will flow in the direction that the arrow is pointing.

TESTING A DIODE: An ohmmeter is used for testing a diode. When the ohmmeter leads are correctly connected to the diode terminals a continuity will be indicated in only one direction. When the ohmmeter leads are reversed there will be no continuity indicated. When continuity is indicated in both directions the diode has shorted and must be replaced. When no continuity is indicated in either direction, the diode is open and must be replaced. A simple two-step procedure is used to test a diode, as follows:

1. Set the ohmmeter on the 1X scale and zero the ohmmeter. Connect the ohmmeter leads to the diode. Determine if there is continuity through the diode. See Figure 81-8.

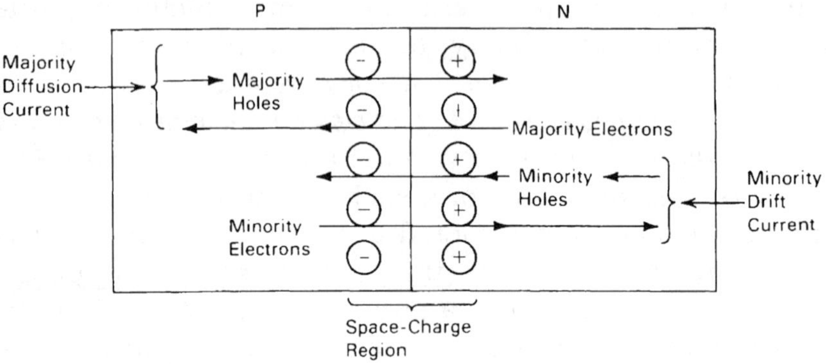

P   N

Majority
Diffusion → Majority Holes

Majority Electrons

Minority Holes

Minority Drift Current

Minority Electrons

Space-Charge Region

Figure 81-8 Currents across a P-N junction with no external connections.

2. Reverse the ohmmeter leads on the diode. See Figure 81-9.

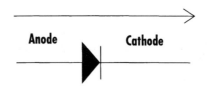

Anode   **Cathode**

Figure 81-9 Schematic symbol of a diode.

Determine if there is continuity through the diode. There should be continuity indicated in only one direction.

IDENTIFYING DIODE LEADS: When replacing a diode, the leads must be identified before the installation. The leads are connected to the anode and the cathode. There are several different case styles used in the manufacture of diodes that require different methods of identifying the leads. The larger diodes usually have the diode symbol printed or stamped on the case identifying the leads. The smaller diodes usually have a line of band printed around the diode case. This line is used to properly identify the leads. This printed line represents the line drawn in front of the arrow on the schematic symbol in the diagram. See Figure 81-10.

It should be remembered that an ohmmeter can be used

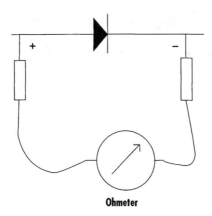

Ohmeter

Figure 81-10 Testing a diode.

to determine diode lead identification. However, the polarity of the ohmmeter must be known. To make the diode forward biased, the positive lead of the ohmmeter is connected to the anode and the negative lead is connected to the cathode.

Unit 82: The Light Emitting Diode (LED)

The LED is relatively inexpensive and is becoming more popular in electronic control circuits for air conditioning, heating, and refrigeration systems. There is no filament to burn out requiring replacement of the LED. They are used as pilot lights as indicators and on numerical displays to indicate system operation. They are used on electronic thermostats to indicate the condition that the system is operating under.

OPERATION: When the standard P-N junction diode absorbs light it can produce an electrical current. When the diode is under a forward-bias condition the light is emitted because of the recombining of the electrons and holes in the junction. The schematic symbol used to indicate an LED is shown in Figure 82-1.

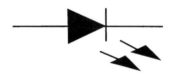

Figure 82-1 LED schematic symbol.

This is the same symbol that is used for the junction diode, with the exception of an additional arrow pointing from the diode. The arrow is an indication that light is being emitted by the diode. A photo diode is another type of control that is turned when it is subjected to light. The photo diode schematic symbol is shown in Figure 82-2.

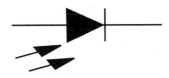

Figure 82-2 Photodiode schematic symbol.

It must be noted that the direction of the arrow is in the opposite direction to that of the LED. This direction of the arrow indicates that the diode must receive light for its operation.

The LED and the junction diode are very much alike in many ways. Both of them are rectifiers and will allow current to flow in only one direction. The major difference is that the LED has a higher voltage drop across it than the junction diode. A silicone junction diode requires about 0.7 volts for it to turn on. The LED requires about 1.7 volts to turn it on. The LED is more difficult to test with an ohmmeter because of this higher voltage drop, because the ohmmeter must provide the required voltage to turn the LED on. Most ohmmeters are not capable of producing this high a voltage. To test an LED it is usually easier to build a circuit including the LED and see if it will operate as desired.

The LED, when used in a circuit, usually requires about 20 milliamps (ma) or less. As an example, if the LED is to be used in a 12 volt DC electrical circuit, a current-limiting must also be installed in the circuit to protect the LED from high voltage. See Figure 82-3.

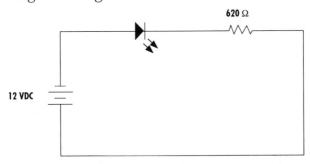

Figure 82-3 Current limiting resistor in series with an LED.

To determine the size of resistor needed use the following steps:

$$R = E/R$$
$$R = 12/0.020$$
$$r = 600 \text{ ohms } (\Omega)$$

Use the closest rated resistor for this application. It may be better to use one of higher rating than one with a lower rating to prevent damage to the LED.

It is necessary to know which terminal is the anode and which is the cathode. To determine this, hold the diode with the leads towards you. There is one flat side on the diode which is close to one of the leads. See Figure 82-4.

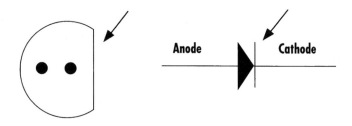

Figure 82-4 Identifying diode leads.

This flat side represents the line in front of the arrow on the diode symbol.

Unit 83: The Transistor

Transistors are two-junction three-terminal devices that operate very much like the P-N junction diode. Transistors have taken the place of the vacuum and gas tubes that were used in many applications, such as; the radio. Transistors have many advantages over the tubes previously used. Some of these advantages are: **(1)** very low operating voltages are used, **(2)** no heating filament is required, **(3)** power consumption is low, resulting in better circuit efficiency; **(4)** they are smaller in size; **(5)** they are virtually shock proof; and **(6)** they are long lived and have essentially no aging effects.

An NPN transistor is basically made up of an emitter, a base, and a collector. See Figure 83-1.

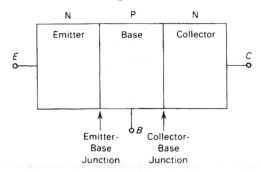

Figure 83-1 NPN transistor structure.

The two regions on the outside are doped to make them the N-type and the center region is doped to make it a P-type. Doping is the process of adding impurities to a semiconductor. The center region **(P)** is known as the base, the first N region is known as the emitter, and the other N region is known as the collector. In these devices there are two P-N junctions; **(1)** the emitter-base **(E-B)** junction, which is formed by the junction of the emitter and base regions, and **(2)** the collector-base **(C-B)** junction, formed by the junction of the collector and base regions. There is an external connection fixed to each of these regions, making the transistor a three-terminal device.

A similar type of structure is the PNP junction transistor. See Figure 83-2.

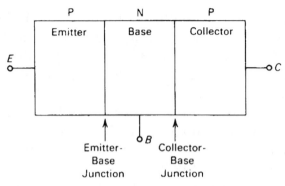

Figure 83-2 PNP transistor structure.

In these types of transistors the doping is exactly opposite to that used in NPN types.

Both of these types of transistors, PNP and NPN, operate in essentially the same way except for the differences in polarity of the current and voltage. Thus, our discussion will be concentrated on the NPN type with periodical references to the PNP type.

TRANSISTOR OPERATION:

Transistors can be biased in four different ways. See Table 83-1.

Biasing transistor junctions

Condition	E-B junction	C-B junction	Region of operation
I	Forward biased	Reverse (or un-) biased	Active
II	Forward biased	Forward biased	Saturation
III	Reverse biased	Reverse (or un-) biased	Cutoff
IV	Reverse biased	Forward biased	Inverted

Table 83-1 Biasing transistor junctions.

Out of these four possibilities we will discuss only condition I. In this condition the E-B junction is forward biased and the C-B junction is reverse (or un-) biased. Operating in this condition is referred to as *active operation*. See Figure 83-3.

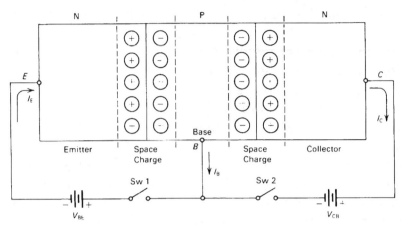

Figure 83-3 Biasing the NPN transistor for active operation.

In this circuit the battery labeled V_{BE} produces a forward bias on the battery labeled V_{CB} produces a reverse bias on the C-B junction. The current that flows through each terminal is labeled I_E (the emitter junction), I_B (the base current) and C_I (the collector current). The indicated directions are assumed to be in the positive direction for an NPN transistor. The E-B junction will be forward biased when only switch 1 is closed. We can safely assume that a comparatively large current flow will cross the E-B junction because of the base and emitter regions are the same as the P-N junction diode. This current will be made up of majority carriers diffusing across the junction from the base to the emitter. See Figure 83-4.

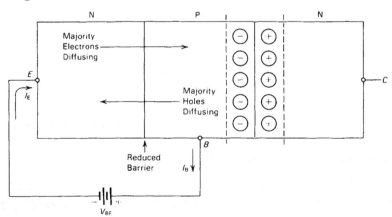

Figure 83-4 With only switch 1 closed.

The total flow of current crossing the E-B junction equals the sum of the electron diffusion current and hole diffusion current. The base region in junction transistors is deliberately doped very lightly compared to the emitter region. This light doping allows the number of electrons diffusing from the emitter to amount to more than 90% of the total current flow. This light doping of the base also allows many of the electrons from the emitter to pass through the base region to the positive battery terminal without merging with a hole and recombining. Only a very few of these diffusing electrons will combine with a hole in the base region. Note that large amounts of equal currents flow into the emitter and out of the base region. (I_E = I_B) and that no current is flowing in the collector (I_C = 0). See Figure 83-4.

When only switch 2 is closed, the junction C-B will be reversed biased. See Figure 83-3. When a reversed bias condition exists, any current flow across the junction C-B will be reversed leakage current that has been set in motion by the minority carriers, which have received enough heat energy to move and are moved by the potential barrier. This is a relatively small current and is totally dependent on the temperature. See Figure 83-5.

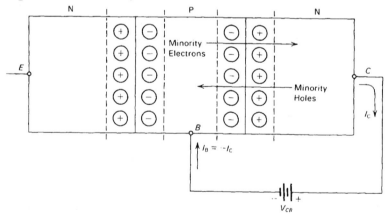

Figure 83-5 Switch 2 closed and switch 1 open.

Note that the current will actually flow into the base terminal. Because of this the current I_B will be negative, as it actually does flow opposite to the assumed current flow direction. When this condition occurs, very small currents flow from the collector and into the base ($I_C = I_B$). No emitter current flows at this time ($I_E = 0$).

When both switch 1 and switch 2 are closed, the emitter current (I_E) is rather large, the base current (I_B) is very small, and the collector current (I_C) is a large current. See Figure 83-6.

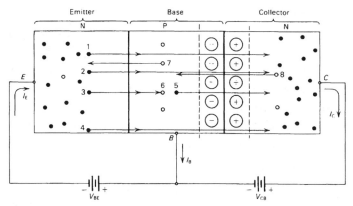

Figure 83-6 Both switch 1 and switch 2 closed; the NPN transistor operating in the active region.

In this figure some of the holes and some of the electrons are numbered to aid in understanding the description of transistor action.

In Figure 83-6, the junction E-B is forward biased because of the voltage V_{BE}. The majority carriers will diffuse across junction E-B as indicated by the electrons 1, 2, 3, and 4 from the emitter and hole 7 from the base emitter. The base, in this case, is very lightly doped and most of the majority current is produced by the electron movement, while very little is carried by the holes. Junction C-B is still reverse biased because of voltage V_{CB}. Therefore, only the minority carriers cross the junction C-B, as indicated by electron 5 traveling from the base to the collector and hole 8 traveling from the collector to the base. These minority carriers gain their energy from the ambient heat and their current is essentially I_{CBO}, discussed earlier. There are also minority electrons found in the base region which have diffused from the emitter. They are electrons 1 through 4, which are called minority carriers upon entering the base. These electrons diffuse toward junction C-B through the base region, where they are attracted into the collector by the potential barrier. Electrons 1, 2, and 4 continue through to the collector. When these electrons cross junction C-B they add to the existing reverse leakage current I_{CBO} and produce an increased collector current. However, electron 3 does not pass through the base to the collector because in the base region it recombines with hole 6. The base region is made deliberately narrow (approximately 0.001 in.) to aid the electrons in diffusing from the emitter to the collector. In a good transistor only a very few of the electrons recombine with a hole in the base.

The NPN and the PNP transistors both have their own electronic symbols. See Figure 83-7.

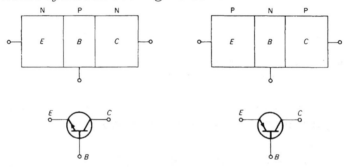

Figure 83-7 Transistor symbols.

An arrow is used to indicate the direction of emitter current flow when junction E-B is reverse biased. If the emitter were forward biased, the arrow would be reversed.

One of the transistor terminals is chosen as the common terminal, one is chosen as the input terminal, and one is chosen as the output terminal. The common-base configuration indicates the base as the common terminal, the emitter or the input terminal, and the collector as the output terminal. See Figure 83-8.

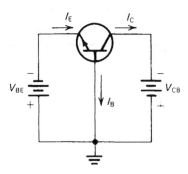

Figure 83-8 Common-base configuration for NPN transistors.

With the biased voltages applied, the junction E-B is forward biased, producing the emitter current (I_E) and the base current (I_B) as indicated. Junction C-B is reverse biased, producing the collector current (I_C) as shown. The emitter terminal is considered to be the input terminal in the common-base configuration. The emitter current is the input current and the collector current is the output current. For any given emitter current there will be a corresponding collector current output.

IDENTIFYING TRANSISTOR TERMINALS: Usually, the style of the transistor case will allow the leads to readily identified. The TO5, the TO18, and the TO 3 cases are prime examples of this type. The leads of the TO5 or the TO18 transistors can be easily identified by holding the transistor with the leads facing you. See Figure 83-9.

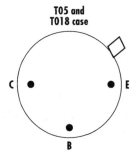

Figure 83-9 Identifying terminals on TO5 and TO18 transistors.

The metal tab on the transistor case is nearest the emitter terminal. The base and collector terminals are positioned as shown in the figure.

The TO3 type transistor can be readily identified by holding the the terminals facing you and down. The emitter terminal is on the left and the base terminal is on the right side of the transistor. The case acts as the collector for this type transistor case. See Figure 83-10.

TO 3 case

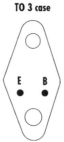

Figure 83-10 Identifying terminals on TO3 transistors.

TESTING THE TRANSISTOR: An ohmmeter is generally used to test a transistor. The transistor can be identified as NPN or PNP when the polarity of the ohmmeter leads is known. To an ohmmeter, an NPN transistor will act like two diodes with their anodes connected together. When the positive lead of the ohmmeter is connected to the base terminal of the transistor, a diode junction should be indicated between the base-collector and the base emitter. When the negative ohmmeter lead is connected to the base terminal of an NPN transistor, no continuity should be indicated between the base-collector and the base-emitter junction on the ohmmeter.

A PNP transistor will act like two diodes with their cathodes connected together. When the negative lead of the ohmmeter is connected to the base terminal, a diode junction should be indicated between the base-collector and the base-emitter. When the positive lead of the ohmmeter is connected to the base lead there should be no continuity indicated between the base-collector or the base-emitter.

The following procedure is one way to test the condition of a transistor:

1. Determine the polarity of the ohmmeter leads. A diode may be used to make this determination. Continuity through the diode will be indicated only when the positive ohmmeter lead is connected to the anode and the negative lead is connected to the cathode. See Figure 83-11.

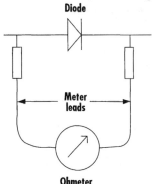

Figure 83-11 Determining the polarity of the Ohmmeter leads.

2. When testing an NPN transistor, the positive ohmmeter lead is connected to the base terminal and the negative lead is connected to the collector terminal. If the transistor is good, continuity should be indicated on the ohmmeter. Approximately the same reading should be indicated as that found when the diode was tested. See Figure 83-12.

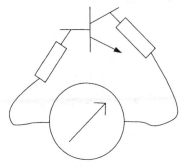

Figure 83-12 Testing an NPN transistor.

3. Leave the positive lead connected to the base terminal of the transistor and connect the negative lead to to the emitter terminal of the transistor. The ohmmeter should indicate a forward diode junction. See Figure 83-13.

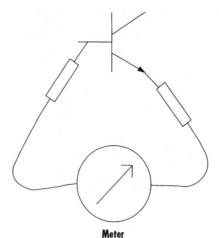

Meter

Figure 83-13 Testing an NPN transistor.

When the ohmmeter does not indicate continuity between the base- collector terminals or the base-emitter terminals the transistor is open.

4. When the negative ohmmeter lead is connected to the base lead and the positive lead is connected to the collector, the ohmmeter should indicate infinity or no continuity. See Figure 83-14.

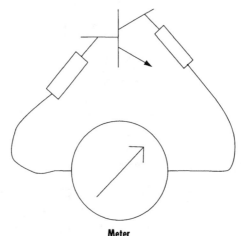

Meter

Figure 83-14 Checking the base to collector of an NPN transistor.

5. Leave the negative ohmmeter lead connected to the base terminal and move the positive lead to the emitter terminal. There should not be any continuity indicated. See Figure 83-15.

Ohmeter

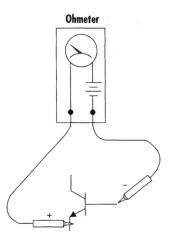

Figure 83-15 Checking the base to emitter of an NPN transistor.

However, if a very high resistance is indicated by the ohmmeter, the transistor has an internal leak. It may still function in the circuit. The transistor is shorted when a low resistance is indicated on the ohmmeter.

6. When testing a PNP transistor, the ohmmeter leads must be reversed before the test can be properly accomplished. If the negative ohmmeter lead is connected to the base lead and the positive lead is connected to the collector or the emitter lead, a forward diode junction should be indicated. See Figure 83-16.

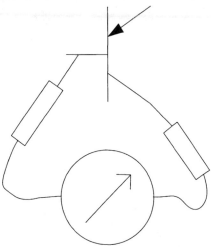

Meter

Figure 83-16 Testing a PNP transistor.

7. When the positive ohmmeter lead is connected to the base lead and the negative lead is connected to the collector or emitter terminal there should be no continuity indicated on a PNP transistor. See Figure 83-17.

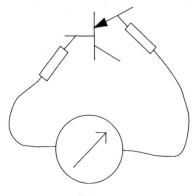

Meter

Figure 83-17 Testing a PNP transistor.

Unit 84: The Unijunction Transistor (UJT)

The unijunction transistor has two bases and one emitter. It is considered to be a digital device because it is either completely on or completely off.

They are manufactured by combining three layers of semi-conductor materials. See Figure 84-1.

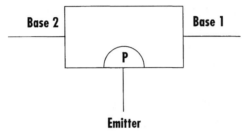

Figure 84-1 A unijunction transistor.

The schematic symbol for the UJT is shown in Figure 84-2.

Figure 84-2 UJT schematic symbol.

The base diagram and terminal location are shown in Figure 84-3.

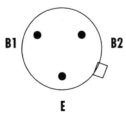

Figure 84-3 UJT base diagram.

Notice that the metal tab is located between the base 2 terminal and the emitter terminal.

CHARACTERISTICS OF THE UJT: This type of transistor has two paths for the current to flow through. One of the paths is from base 1 (B1) to base 2 (B2). The other path is through B1 and the emitter. When it operating normally there is no current flowing through either path until the applied voltage reaches approximately 10 volts higher than the voltage applied to B 1. The UJT will turn on when the voltage applied to the emitter reaches a point of 10 volts positive higher than the voltage applied to terminal B1. The current then flows through the path from B1 to B2 and from the emitter through B1. The current will continue to flow the UJT until the voltage applied to the emitter drops to approximately 3 volts higher than the voltage applied to B1, the UJT will turn off. It will remain off until the applied voltage to the emitter again reaches approximately 10 volts positive higher than that applied to terminal B1.

OPERATION: UJT transistors are generally used in circuits such as capacitor discharge systems. See Figure 84-4.

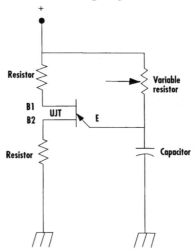

Figure 84-4 UJT operation.

In operation, the variable resistor controls the amount of time for charging the capacitor. When the capacitor charge reaches approximately 10 volts, the UJT will turn on and allow current to flow through it and discharge the capacitor by the circuit through the emitter to B1. When the capacitor charge reaches approximately 3 volts, the UJT will turn off, interrupting the circuit and the capacitor begins charging again. When the resistance that is connected in series with the capacitor is changed the charge time of the capacitor is also changed. The opening and closing rate of the UJT can be controlled by varying this resistance.

The UJT can be used to furnish a large output voltage because the output is produced by discharging the capacitor into the circuit. See Figure 84-5.

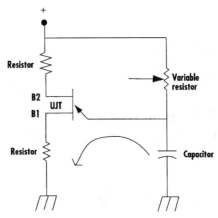

Figure 84-5 Output voltage of a UJT.

The output of the UJT is generally used to trigger an SCR gate.

The rate that the UJT will cycle on and off is determined by the amount of resistance and capacitance connected to the emitter of the UJT. Therefore, the amount of capacitance that can be connected to the UJT has limitations. The maximum capacitance that is connected to a UJT should be limited to about 10μf. When a capacitor that is too large is connected to the UJT, the spike voltage caused by the capacitor discharging could possibly damage the UJT.

TESTING THE UJT: The condition of the UJT can be determined by the use of an ohmmeter. The UJT will act similar to two resistors connected to a common diode. The common junction of the two resistors will seem to be at the UJT emitter terminal. See Figure 84-6.

Figure 84-6 The UJT when tested with an ohmmeter will act like two resistors connected to a diode.

If the positive lead of the ohmmeter is connected to the emitter terminal, a diode junction should be indicated from the emitter terminal to B2 and another diode connection between the emitter and B1. When the negative lead of the ohmmeter is connected to the emitter terminal of the UJT, there should be no continuity between either base terminal and the emitter terminal.

To test a UJT, use the following steps:

1. Determine the positive and negative leads of the ohmmeter. A diode can be used to make this determination. Continuity will be indicated by the ohmmeter when the positive lead is connected to the anode terminal and the negative lead is connected to the cathode terminal. See Figure 84-7.

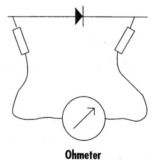

Ohmeter

Figure 84-7 Determining the ohmmeter lead polarity.

2. The positive ohmmeter lead is now connected to the emitter terminal and the negative lead is connected to the B1 terminal. A forward diode junction should be indicated by the ohmmeter. See Figure 84-8.

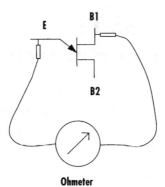

Ohmeter

Figure 84-8 Testing continuity through the E and B1 terminals of a UJT.

3. Leave the positive ohmmeter lead connected to the emitter terminal. Connect the negative lead to B2 terminal. A forward diode junction should be indicated by the ohmmeter. See figure 84-9.

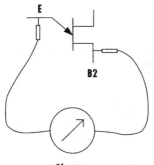

Ohmeter

Figure 84-9 Testing continuity through the E and B2 terminals of a UJT.

4. When the negative ohmmeter lead is connected to the emitter terminal there should be no continuity indicated with the positive lead connected to terminal B2. See figure 84-10.

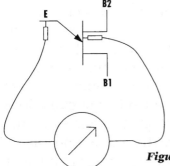

***Figure 84-10 Testing continuity through the
B1 to E terminals of a UJT.***

Ohmeter

Unit 85: The Silicone Controlled Rectifier (SCR)

The SCR is commonly referred to as a PNPN junction because of the way it is made. It has four (4) layers of semiconductor material connected together with three leads extending from it. See Figure 85-1.

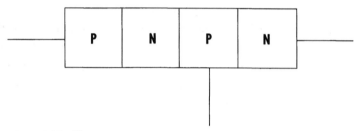

Figure 85-1 SCR diagram.

The symbol used in schematic drawings for the SCR is shown in Figure 85-2.

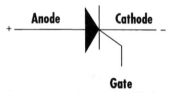

Figure 85-2 SCR symbol.

This symbol is similar to the one used for the diode except that a gate terminal has been added.

CHARACTERISTICS OF AN SCR: Silicon-controlled rectifiers are a part of the thyristor type devices. Thyristors are considered to be digital devices because they are either on or off. Just like a manual light switch. There is no in between. They are especially useful in circuits that an electronic device is used to control a large amount of electric power. When an SCR has been installed in a circuit and is in the off position there will be a voltage which is equal to the supply voltage indicated between the anode and the cathode. There is no current flow through the circuit at this time. When the SCR is turned on, there should be approximately one volt indicated between the anode and the cathode of the SCR. The resistor in the circuit limits the current flow to about 2.4 amps (240 volts ÷ 100Ω = 2.4 amps). The SCR now has 2.4 amps of current flowing at one volt. Therefore it must dissipate 2.4 watts of heat (1 volt x 2.4 amps = 2.4 watts). See Figure 85-3.

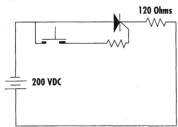

Figure 85-3 An SCR in the off position.

SCRs IN A DC ELECTRIC CIRCUIT: When an SCR is connected into a DC circuit, power on the gate will cause the SCR to turn on but it will not turn it off. The gate is connected with the same electrical polarity as the anode-cathode section of the SCR. See Figure 85-4.

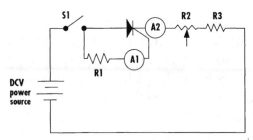

Figure 85-4 Gate voltage on the SCR.

When this connection is made it will turn on the SCR. When the SCR is turned on by the gate voltage, it will remain on until the current flow through the anode-cathode section drops to a value low enough to allow the SCR to turn off.

Holding Current: Holding current is the current that is necessary to cause the SCR to remain turned on once it has been turned on. When an adjustable resistor **(R1)** is installed in the circuit and adjusted to 0 resistance and switch one is turned on, there will be no current flow through the anode-cathode section of the SCR and the SCR will not be triggered. Should the value of the adjustable resistor **(R1)** be slowly decreased the current flow through the gate-cathode circuit will be slowly increased. When the amount of current flowing through the gate reaches the turn on value the SCR will be turned on. When the SCR turns on, current will be flowing through the anode-cathode circuit of the SCR, it will be turned on. At this point the voltage drop across the SCR will be about one volt. When the SCR is turned on the gate has no further control over the operation of the SCR. In fact, it could be disconnected from the lead without any affect on the operation of the SCR. When the SCR turns on, the anode-cathode acts like a short circuit and the current flow through it is limited only by resistor R3. At this point, if the value of the resistance of the variable resistor **(R2)** is slowly increased the current flow through the anode-cathode circuit will slowly decrease until the drop-out current is reached. At this point the SCR will open and no current will flow through it.

SCRs IN AN AC ELECTRIC CIRCUIT: When an SCR is connected into an AC electric circuit it operates as a rectifier. That is, the output from it will be a DC voltage. An SCR that is used in an AC circuit acts almost like it would if it were connected into a DC circuit, except the change in operation when the waveform drops its peak positive value to its 0 value at the end of each half cycle of the current. When the 0 value is reached the SCR is allowed to turn off. This requires that the gate re-trigger the SCR at the end of every positive half cycle of the current. See Figure 85-5.

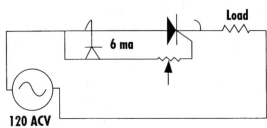

Figure 85-5 *In an AC circuit, the SCR turns on when the AC wave form reaches its highest value.*

If we were to install a variable resistor in the lead to the gate connection of the SCR and adjust it to allow 10 ma of current to flow just as the value of the applied voltage reaches its peak value, the SCR will turn on. At this point current will start flowing through the load when the AC waveform is at its peak positive value. The current will flow through the load resistor until the drop in the sine wave causes the current to drop below the current holding value of the SCR, at which time the SCR will turn off and stop the flow of current. The SCR will not turn back on until the AC waveform reaches the positive peak value of the waveform. The SCR will not be turned on when the electrical cycle enters the negative half of the waveform because it is reverse biased and cannot be turned on with a negative signal.

If the value of the resistance in the gate lead is reduced, a current of 10 ma will be reached before the AC waveform reaches it positive peak value. See Figure 85-6.

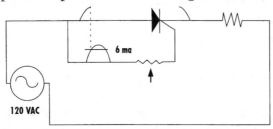

Figure 85-6 Triggering on the SCR before the wave form reaches its peak value.

Thus, the SCR can be turned on earlier than when a higher ma is used, and current is allowed to flow through the load for a longer period of time. This action allows for a higher average voltage drop across the load resistance. For each drop in value of resistance in the gate lead the SCR will be turned on earlier in the cycle. See Figure 85-7.

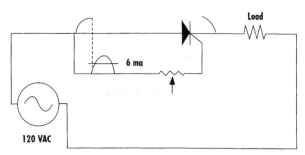

Figure 85-7 Lower resistance in the gate lead allows the SCR to turn on earlier.

Again, this allows the current to start flowing through the circuit earlier than before and allow a higher average voltage drop across the load resistance.

Using the SCR is a circuit of this type allows the SCR to control only the first part of the positive waveform. The SCR must be fired before the waveform reaches 90, or its peak value.

TESTING THE SCR: As with troubleshooting any piece of equipment, the first thing is to give a complete and thorough visual inspection of the equipment. Inspect the printed circuit boards for loose connections, shorted circuits, and for open circuits. A thorough visual inspection will usually save many hours of troubleshooting.

An ohmmeter can be used to test an SCR. To make the test, zero the ohmmeter, then connect the positive lead of the ohmmeter to the anode connection and the negative ohmmeter lead to the cathode connection. There should be continuity indicated between these two points. If the ohmmeter lead to the gate lead is removed there may be or may not be conduction through these SCR. This will depend on the current output of the ohmmeter. If it is providing enough current to keep the current flow above the holding current value of the SCR continuity will be indicated. If continuity is indicated by the ohmmeter before the gate is connected to the ohmmeter, the SCR is shorted and must be replaced. If there is no continuity indicated through the SCR after the gate is connected to the ohmmeter, the SCR has an open circuit.

Use the following steps to test an SCR:

1. Determine which ohmmeter lead is positive and which is negative by using a junction diode. Continuity will be indicated only when the positive ohmmeter lead is connected to the anode or the diode and the negative ohmmeter lead is connected to the cathode. See Figure 85-8.

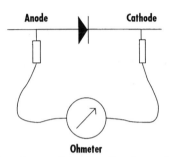

Ohmeter

Figure 85-8 Determining the positive leads of an ohmmeter.

2. Connect the negative ohmmeter lead to the cathode of the SCR and the positive ohmmeter lead to the anode. There should be no continuity indicated by the ohmmeter. See Figure 85-9.

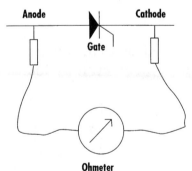

Ohmeter

Figure 85-9 Testing the circuit between the anode and the cathode of an SCR.

3. Connect the gate of the SCR to the anode with a jumper wire. A forward diode junction should be indicated. When the jumper wire is removed the SCR may remain on or it may turn off, depending on whether the ohmmeter is providing sufficient current to keep the SCR above its holding current requirement. See Figure 85-10.

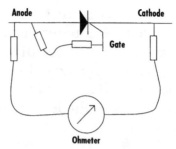

***Figure 85-10** Testing the circuit between the gate and the anode of an SCR.*

4. Connect the positive lead of the ohmmeter to the cathode connection and the ohmmeter negative lead to the anode of the SCR. No continuity should be indicated by the ohmmeter. See Figure 85-11.

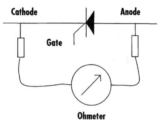

***Figure 85-11** Testing the reverse flow of current through the cathode to anode circuit.*

5. Connect the gate lead to the anode lead with a jumper wire. There should be no continuity indicated on the ohmmeter. See Figure 85-12.

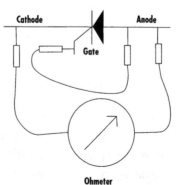

***Figure 85-12** Testing the ?SCR with a jumper wire between the gate and anode leads with a negative ohmmeter lead to anode and the positive to the cathode.*

Remember that SCRs that are designed for switching large current loads may indicate a small amount of current leakage when this test is used. This is usually a normal condition for these types of devices.

Unit 86: The Triac

This semiconductor device acts like two SCRs connected in reverse parallel. Because of this they can be used for controlling an AC current load. They are two PNPN junctions connected in parallel and have three connections. See Figure 86-1.

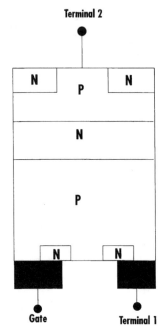

Figure 86-1 TRIAC construction.

The two gates are connected together which allows only three connections to the outside of the device. See Figure 86-2.

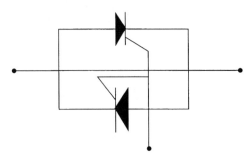

Figure 86-2 TRIAC connections.

In operation, the TRIAC can be caused to conduct electricity in both directions by a gate signal that is similar required by the SCR. They are mostly used in low current applications. Thus, they cannot be used to replace an SCR which is used in high current applications. It should be noted that when an SCR is connected into a circuit its output current will be DC. When it is connected into an AC circuit its output current will be AC. It will conduct the complete AC current cycle (both negative and positive sides of the sine wave). The Schematic symbol for a TRIAC includes a gate (G) and two terminals (T1 and T2). See Figure 86-3.

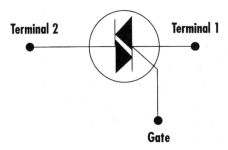

Terminal 2   **Terminal 1**

Gate

Figure 86-3 TRIAC schematic symbol.

TRIAC OPERATION: The TRIAC represents a high resistance to the flow of current and will block its flow in either direction until the correct triggering signal is sensed by the gate. When the gate senses the correct signal, current will be conducted in either direction through the device. The direction depends on the type of signal the gate senses. The major difference between the TRIAC and an SCR is the polarity of the signal sensed by the gate. There are four conditions that will trigger the TRIAC, as follows:

1. T2 positive; Vg positive
2. T2 positive; Vg negative
3. T2 negative; Vg positive
4. T2 negative; Vg negative

The most popular conditions are numbers 1 and 4 because these conditions require less gate current to trigger the TRIAC.

When the TRIAC has been turned on in either direction, it operates much like an SCR because the voltage will drop to a very low value before it is turned off. That is, the TRIAC will remain on until the holding current is reduced to a very low value. Also, the current sensed by the gate cannot turn off the TRIAC. It will turn off only when the holding current drops below the hold in value.

The TRIAC, like the SCR is either on or off. Thus, there is no heat to be dissipated with the associated loss in power. The TRIAC has no reverse breakdown of high voltage and current because when the voltage reaches the prescribed value the TRIAC will automatically turn on and allow this high current to flow through.

TRIAC APPLICATIONS: TRIACS are becoming very popular in the starting and operation of electric motors. They are gaining acceptance as replacements for the starting relays used to start single-phase compressor motors. Their popularity is due to the fact that they operate much faster and are more trouble free than the electro-mechanical type being replaced.

When the compressor-motor is first started there is a very high current flow through both windings of the motor. High current flow through a TRIAC causes it to turn on and allow the high current to pass through it. When the motor approaches operating speed, the current draw of the motor drops because of the counter-EMF build-up in the windings. When this current draw drops below the holding current of the TRIAC it is turned off, stopping the flow of current through it. At this point, the starting winding of the motor is disconnected from the motor operating circuit. See Figure 86-4.

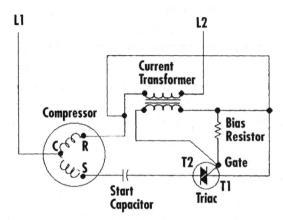

Figure 86-4 TRIAC connection when used as a motor starting device.

TESTING THE TRIAC: A TRIAC circuit and an ohm-meter can be used to test a TRIAC. Use the following procedures when making this test: See Figure 86-5.

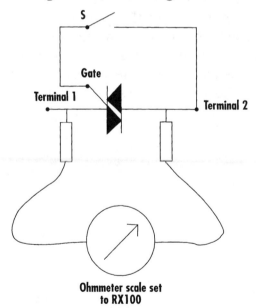

Ohmmeter scale set
to RX100

Figure 86-5 Testing the TRIAC using an ohmmeter.

The TRIAC should be replaced should either one of these tests fail:

1. Set the ohmmeter on the R x 100 scale.

2. Zero the ohmmeter for that scale.

3. Connect the negative lead of the ohmmeter to terminal T1 of the TRIAC.

4. Connect the positive ohmmeter lead to terminal T2 of the TRIAC. The ohmmeter should indicate an infinite resistance through this circuit. The minimum resistance indicated through this circuit is 250,000 ohms.

5. Complete a short of the gate by closing switch S. The ohmmeter should now indicate a resistance of between 10 and 50 ohms. On some ohmmeters a reading this small may be difficult to read on the R x 100 scale.

6. Now, open the switch. The ohmmeter should still indicate that there is between 10 and 50 ohms resistance through this circuit.

7. Reverse the ohmmeter leads. Connect the positive lead to T1 and the negative lead to T2. An infinity resistance should be indicated on the ohmmeter.

8. Complete a short of the gate by closing the switch S. The resistance reading on the ohmmeter should now be so small that it is almost impossible to read.

9. Open switch S. The resistance reading should not change. It should be the same as that indicated in step 8.

Unit 87: The Diac

The DIAC is a three-layer diode that will allow current to pass in either direction. Each of the junctions are heavily and evenly doped so that they are just alike. See Figure 87-1.

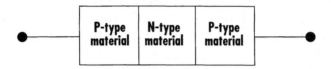

Figure 87-1 Construction of a DIAC.

In operation, they are used to switch voltages up to a few hundred volts. The DIAC looks much like any electrical resistor used in circuits. See Figure 87-2.

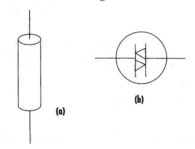

Figure 87-2 DIAC casing (a) and symbol (b).

DIAC OPERATION: Just like the TRIAC, the DIAC will allow voltage to pass in either direction. Because of this characteristic, it operates very similar to two zener diodes connected in electrical series and in opposite directions. The operation of a DIAC is similar to a breakover device. See Figure 87-3.

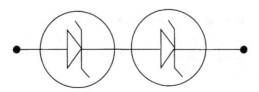

Figure 87-3 A DIAC represented by two zener diodes.

During operation, as the applied voltage is increased, but remains below the breakover voltage, the amount of current passing through the DIAC will be reduced. Because of this, it is often said that the DIAC has a negative resistance.

DIAC APPLICATIONS: Because DIACs operate as breakover devices, they are often used in the gate control circuits of TRIACS to improve gate operation. When used in this manner an advantage of applying a pulse of gate current instead of applying a sinusoidal current to the gate terminal of the device being controlled. This pulse current to the gate provides a better control of the operating sequence of the TRIAC.

TESTING A DIAC: There are basically two methods that can be used to test a DIAC: the use of an oscilloscope and the use of an ohmmeter. The best procedure is to use an oscilloscope. Use the following steps to test a DIAC with an oscilloscope:

1. Construct a test circuit. See Figure 87-4.

2. The circuit is now connected to a power source. Turn on the power.

3. Adjust the oscilloscope using the manufacturer's instructions.

4. Connect the oscilloscope to the circuit. See Figure 87-4.

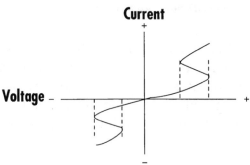

Figure 87-4 Testing a DIAC using an oscilloscope.

5. When a uniform trace is indicated on the oscilloscope, the DIAC is good. See Figure 87-5.

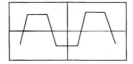

Figure 87-5 Oscilloscope trace of a good DIAC.

A trace other than a uniform one indicates that there is trouble with the DIAC. It should be replaced.

The ohmmeter can be used to test a DIAC for an open circuit. It should be noted that an ohmmeter test is not as accurate as the test made with the oscilloscope. To test a DIAC with an ohmmeter use the following steps:

1. Set the ohmmeter scale on R x 100 and zero the ohmmeter.

2. Remove one of the DIAC leads from the circuit.

3. Touch the ohmmeter leads to the terminals on the DIAC. See Figure 87-6.

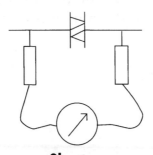

Ohmeter
low resistance reading

Figure 87-6 Testing a DIAC with an ohmmeter.

Record this resistance reading for future reference.

4. To test the DIAC in the opposite direction, reverse the ohmmeter leads on the DIAC terminals. See Figure 87-7.

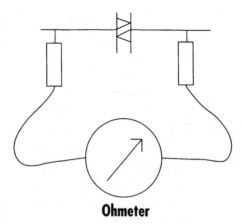

Ohmeter

Figure 87-7 Testing the reverse resistance of a DIAC with an ohmmeter.

These resistance readings should both be low. The low readings are the result of the DIAC acting as two diodes connected in parallel. When the resistance readings found in steps 3 and 4 are slightly different, or if the DIAC is found to be open, test the DIAC with an oscilloscope to verify its condition.

Section 9: Basic Electronic Controls

INTRODUCTION:

This section will describe some of the more popular solid-state controls used in the air conditioning, refrigeration, and heating industry. A general description, operation, some common uses, and applications will be discussed.

Many of the solid-state controls used in air conditioning, and refrigeration systems today are single-function devices. This means that the control has only one function in the operation of the system. They either control or protect only one function in the total systems operation. Some good examples of this type of control is the defrost control used on modern heat pump systems, time-delay devices to prevent short-cycling of a motor, for electric motor protection, they are used sometimes as starting relays for single-phase motors. We will cover the more popular solid-state controls used on these types of systems. There could possibly be several volumes written on solid state controls and systems used in the air conditioning and refrigeration industry.

Unit 88: Solid-State Timers

The solid-state timer is very popular in air conditioning and refrigeration systems. They are popular because they are available in a large variety of types purposes. They are more flexible than the conventional type time-delay relay, but they may be used to perform the same functions. Timers have many functions in this industry; such as, staggering the starting load of electric resistance heaters to prevent an overload on the electrical system, to prevent short-cycling of compressors, and to stager the starting load of large equipment. The equipment manufacturers generally use a set amount of time delay to suit their equipment operation. Adjustable models are also available for field installation to provide some specific function in the operation of the equipment. See Figure 88-1.

Figure 88-1 Solid-state adjustable timer. (Courtesy of International Controls and Measurements.)

Some models have a wide range of adjustment time and some have very small range adjustment. The type chosen will be determined by the function being controlled.

Unit 89: Anti-Short-Cycling Controls

This type of control is probably the most popular used in the air conditioning and refrigeration industry. Their purpose is to prevent the compressor from starting for a predetermined period of time from when it stopped operation. For example, when a pressure control stops compressor operation, the anti-short-cycling control will prevent it from restarting until the time period has passed. They can delay compressor starting from 30 seconds to five minutes, depending on the desires of the equipment manufacturer and the conditions in which it is operating. See Figure 89-1.

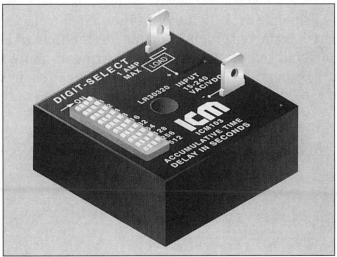

Figure 89-1 Anti- short-cycling control. (Courtesy of International Controls and Measurements.)

If the anti-short-cycling control is not providing the desired amount of delay between operating cycles then the control is not doing its job and should be checked out and probably adjusted or replaced.

The wiring diagram for a typical anti-short-cycling control is shown in Figure 89-2.

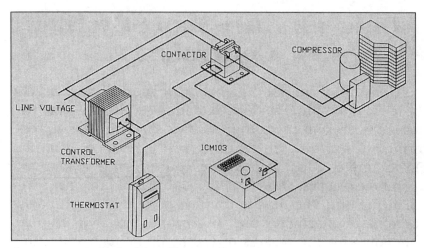

Figure 89-2 Typical wiring diagram for an anti-short-cycling control. (Courtesy of International Controls and Measurements.)

The principle on which these types of controls operate is that they have a very high resistance to the flow of electricity when cold. As the PTC material warms up more voltage can pass through it. When the passed voltage reaches a predetermined value then the controlled device is energized.

Unit 90: Electronic Thermostat

Electronic thermostats are becoming very popular because they can be programmed to control the air conditioning equipment for several functions and at various times during the day.

This type of thermostat uses a microcomputer and a solid-state sensor for precise temperature control. They will operate on most 24 vac heating only or heating and cooling systems. These thermostats incorporate a separate set-back programming for seven independent days. Four separate time-temperature settings per 24 hour period. They usually include a battery back-up in case of a power failure. Some offer short-cycle protection for the compressor. The temperature range is usually from 40 to 90°F. When installing, servicing or adjusting these types of thermostats, be sure to use the correct instructions for the make and model being serviced.

Unit 91: Solid-State Defrost Controls

One of the more popular types is the CMC defrost control that is manufactured by Hamilton Standard. See Figure 91-1.

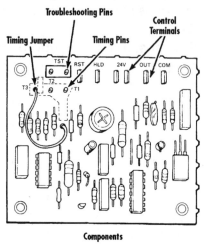

Figure 91-1 Solid state defrost control. (Courtesy of Lennox Industries.)

This type of control provides automatic switching from the normal heating mode to the defrost mode and then back to the heating mode.

These types of controls uses a solid state timer to switch an external defrost relay through 1/4 inch male spade terminals that are mounted on the control circuit board. The defrost control energizes the defrost relay at regular timed intervals. The specific wiring and external components used by different manufacturers will have some affect on the operation of the control. Be sure to check the specific unit wiring diagrams for the correct connecting procedures.

DEFROST CONTROL COMPONENTS:

1. "24V" Terminal

The "24V" terminal receives 24VAC from the control transformer. This terminal provides electrical power to the internal timer and relays. The "24V" terminal must have power at all times so that a "HOLD" is provided between the demands of the thermostat.

2. "COM" Terminal

The "COM" terminal provides the 24VAC common connection.

3. "HLD" Terminal

The "HLD" terminal holds the internal timer in place between the demands from the thermostat and allows the timer to continue timing when the thermostat again demands operation. Generally the "HLD" terminal is connected directly to the thermostat demand.

4. "OUT" Terminal

The "OUT" terminal controls the unit defrost cycle when it is connected to one side of the defrost relay. An internal relay that is connected to the "OUT" terminal closes to allow the defrost relay to be energized and initiate the defrost cycle. At the end of the defrost cycle, the internal relay that is connected to the "OUT" terminal opens to de-energize the external defrost relay.

5. "RST" Terminal

The "RST" terminal resets the internal timer when the power is removed and begins operation of the timer when the power is returned to the terminal. The "RST" terminal is connected to the "COM" terminal through a set of normally closed (NC) contacts in the defrost relay. When the defrost relay contacts open the "RST" terminal loses its power and the internal timer is reset. The control resumes timing when the defrost relay contacts close.

6. Timing Pins 30(T1), 60(T2), 90(T3)

Each of these pins provides a different timed interval between defrost cycles. A jumper connects the pins to the circuit board terminal W1. Table 91-1 shows the timings of each of these pins. See Figure 91-2.

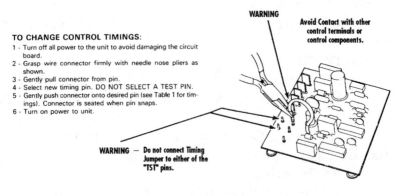

WARNING

Avoid Contact with other control terminals or control components.

TO CHANGE CONTROL TIMINGS:

1 - Turn off all power to the unit to avoid damaging the circuit board.
2 - Grasp wire connector firmly with needle nose pliers as shown.
3 - Gently pull connector from pin.
4 - Select new timing pin. DO NOT SELECT A TEST PIN.
5 - Gently push connector onto desired pin (see Table 1 for timings). Connector is seated when pin snaps.
6 - Turn on power to unit.

WARNING — Do not connect Timing Jumper to either of the "TST" pins.

TABLE 1

DEFROST CONTROL CMC TIMINGS	INTERVAL BETWEEN DEFROSTS JUMPERED CONNECTED TO			DEFROST TIME	
	30 (T1)	60 (T2)	90 (T3)	CHP15, CHP16 SERIES UNITS	HP19 SERIES UNIT
NORMAL OPERATION	30 ± 3 MIN.	60 ± 6 MIN.	90 ± 9 MIN.	10 ± 1 MIN.	14 ± 1.4 MIN.
'TST' PINS JUMPERED TOGETHER	7 ± 0.7 SEC.	14 ± 1.4 SEC.	21 ± 2.1 SEC.	2.3 ± 0.2 SEC.	3.3 ± 0.3 SEC.

Figure 91-2 Defrost cycle timings. (Courtesy of Lennox Industries.)

To change the timing between defrost cycles, simply move the jumper to one of the other available pins.

7. Timing Jumper

A factory installed jumper on the circuit board connects terminal W1 on the circuit board to one of the three timing pins.

8. "TST" Pins

Each circuit board is equipped with a set of test pins to be used when troubleshooting the unit. When they are jumped together, these pins reduce the control timing to about 1/256 of the original time required. See Table 1 in Figure 91-2. To place the control in the test mode refer to Figure 91-3.

TO PLACE CONTROL IN TEST MODE:

1 - Turn off all power to the unit to avoid damaging the circuit board.

2 - Make sure all control terminals are connected as shown on unit wiring diagram before attempting to place control in test mode. See NOTE below.

NOTE — Control will not go into mode when disconnected from unit. Unit load must be applied to control terminals before the control will go into test mode.

3 - Connect jumper to 'TST' pins as shown.

4 - Turn indoor thermostat to heat mode and adjust to highest temperature setting.

5 - Turn on power to unit.

6 - See Table 1 for control timings in 'TST' mode. Follow troubleshooting flow chart to diagnose problems.

7 - Turn on power to unit.

WARNING –

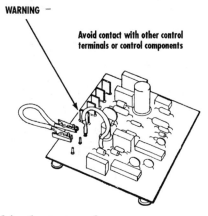

Avoid contact with other control terminals or control components

Figure 91-3 To place the defrost control in the test mode. (Courtesy of Lennox Industries.)

CAUTION: Do not connect the timing jumper to either of the "TST" pins. The "TST" pins must be only jumped together during a test and must not be connected to any other pins. To do so could possibly result in a ruined control.

IMPORTANT: The control will begin the test mode only if a normal load is applied to the control's terminals. Do not attempt to operate or test the control out of the unit.

SEQUENCE OF OPERATION: For this discussion we will use the Lennox CHP 15-510/650 sequence of operation. Refer to Figures 91-4 and 91-5.

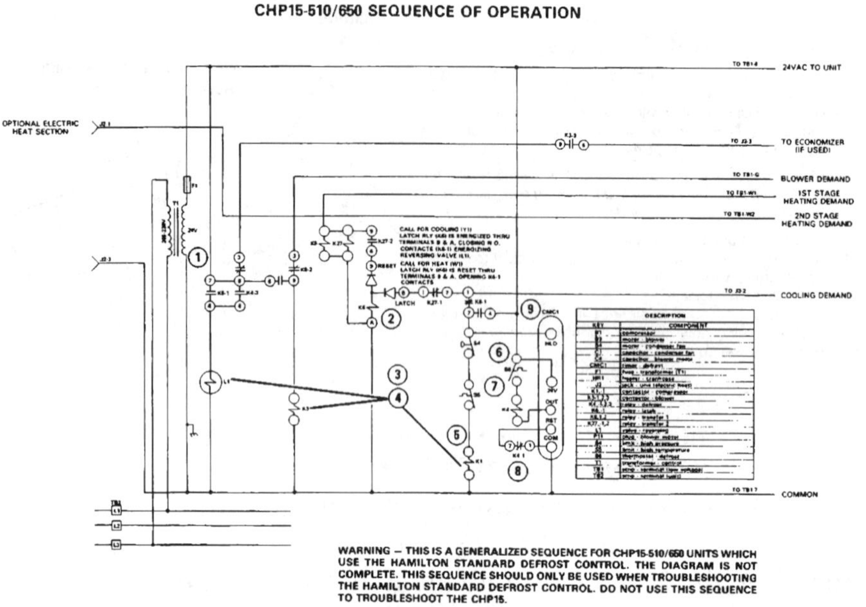

CHP15-510/650 SEQUENCE OF OPERATION

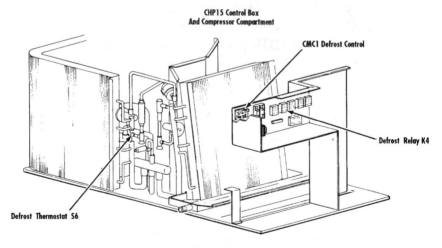

WARNING — THIS IS A GENERALIZED SEQUENCE FOR CHP15-510/650 UNITS WHICH USE THE HAMILTON STANDARD DEFROST CONTROL. THE DIAGRAM IS NOT COMPLETE. THIS SEQUENCE SHOULD ONLY BE USED WHEN TROUBLESHOOTING THE HAMILTON STANDARD DEFROST CONTROL. DO NOT USE THIS SEQUENCE TO TROUBLESHOOT THE CHP15.

Figure 91-4 CHP 15-510/650 sequence of operation.
(Courtesy of Lennox Industries.)

CHP15 Control Box
And Compressor Compartment

CMC1 Defrost Control

Defrost Relay K4

Defrost Thermostat S6

Figure 91-5 Lennox CHP 15 Control box and compressor compartment.
(Courtesy of Lennox Industries.)

1. When line voltage is supplied to terminal T1, 24VAC is supplied to the thermostat and other unit controls.

2. The latch relay K6 controls the operation of the reversing valve L1 during normal operation (contacts K4-3

462

control the reversing valve L1 during the defrost cycle). The latch relay K6 operates as follows:

    a. 24VAC is applied between A and B and closes terminals 7 and 4. Terminals 7 and 4 remain closed when the power is removed.

    b. When 24VAC is applied between terminals A and 9 it causes terminals 7 and 4 to open. Terminals 7 and 4 remain open when the power is removed.

3. When cooling is demanded, power passes through k8-1 contacts 1-7, S4 and S5 to energize the contactor K1. K1 energizes the compressor and the outdoor fan motor. When cooling is demanded the power also passes through K27-1 to energize the latch relay K6 through terminals A and B. K6-1 contacts close to energize the reversing valve L1. Blower demand (TB1-G), is energized simultaneously with the cooling demand and power passes through K8-2 to energize the contactor K3. K3-2 (not shown) switches to energize the blower motor. K3-3 contacts close to energize the economizer.

4. When the cooling demand is satisfied, the contactor K3 drops out. K3-2 contacts open to de-energize the economizer. The latch relay K6 drops out but K6-1 remains closed and the reversing valve L1 remains energized (in case of a subsequent 2nd cooling demand. It keeps L1 from cycling). The reversing valve L1 remains energized until the system is transferred to the heating mode. The contactor K1 drops out and the compressor motor is de-energized.

5. When heating is demanded, relays K8 and K27 are energized. K8-1 switches to energize the contactor K1 through S4 and S5. The compressor motor is energized. K8-2 switches to energize contactor K3. K3-2 (not shown) closes to energize the outdoor blower. K3-3 contacts close to energize the economizer. K27-1 contacts open to allow the latch relay K6 to reset and to prohibit a cooling demand from entering K6. K27-2 contacts close to energize K6 through terminals A and 9. Terminals K6-1 open to de-energize the reversing valve L1.

6. After 30, 60, or 90 minutes of heating demand (as pre-set), the defrost control CMC-1 checks for a need for a defrost cycle by closing an internal relay connected to the "OUT" terminal. The defrost thermostat S6 is closed when the outdoor coil vapor line temperature drops below 35 $\pm$ 4°F. If the defrost thermostat S6 is closed when CMC-1 checks for defrost, relay K4 will be energized to initiate the defrost cycle. K4-1 contacts open to disconnect the econo-mizer and energize the reversing valve L1. The CMC-1 allows K4 to remain energized for a maximum of 10 $\pm$ 1 minutes.

7. The defrost cycle continues until the defrost thermo-stat S6 opens or the 10 minute defrost time is complete. If the heating demand is satisfied during the defrost cycle, 24VAC is still supplied to terminal "24V" and the "OUT" ter-minal, but terminal "HLD" loses its power. This tells the control that the heating demand is satisfied. The control "HOLDS" at that point in the defrost cycle until 24VAC is again applied to the "HLD" terminal (the compressor will start running again at this time).

8. The defrost cycle may be terminated in two ways:

    a. The defrost thermostat S6 opening when the out-door coil vapor line rises above 70 $\pm$ 5°F.

    b. The CMC-1 internal relay connected to the "OUT" terminal opening when the 10 minute defrost cycle is complete.

When either of these two occur, relay K4 is de-energized. K4-1 closes to begin the timer for the next defrost period. K4-2 (not shown) closes to allow the outdoor fan to operate with the compressor. K4-3 switches to energize the econo-mizer and to de-energize the reversing valve L1.

9. If the unit receives a cooling demand while the defrost control CMC-1 is "HOLDING" a defrost, step 3 is repeated. As the outdoor coil warms, the defrost thermostat S6 opens then step 8 is repeated.

INDEX
Electrical Applications For
Air Conditioning And Refrigeration